# Archaeoseismology in the Atalanti Region, Central Mainland Greece

## Theories, methods and practice

Victoria Buck

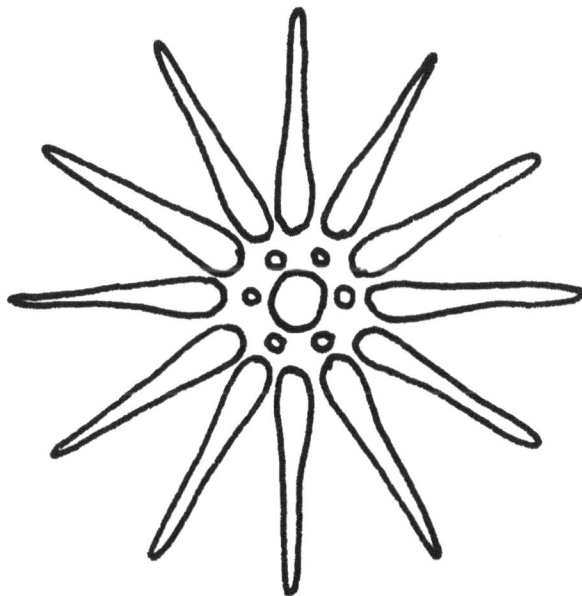

BAR International Series 1552
2006

Published in 2019 by
BAR Publishing, Oxford

BAR International Series 1552

*Archaeoseismology in the Atalanti Region, Central Mainland Greece*

ISBN 9781841719818 paperback
ISBN 9781407330105 e-book

DOI https://doi.org/10.30861/9781841719818

A catalogue record for this book is available from the British Library

This book is available at www.barpublishing.com


BAR Publishing is the trading name of British Archaeological Reports (Oxford) Ltd.
British Archaeological Reports was first incorporated in 1974 to publish the BAR
Series, International and British. In 1992 Hadrian Books Ltd became part of the BAR
group. This volume was originally published by Archaeopress in conjunction with
British Archaeological Reports (Oxford) Ltd / Hadrian Books Ltd, the Series principal
publisher, in 2006. This present volume is published by BAR Publishing, 2019.

# BAR
PUBLISHING

BAR titles are available from:

BAR Publishing
122 Banbury Rd, Oxford, OX2 7BP, UK
EMAIL   info@barpublishing.com
PHONE   +44 (0)1865 310431
FAX   +44 (0)1865 316916
www.barpublishing.com

# Acknowledgements

The primary funding for my research came from a Brunel University (Department of Geography and Earth Sciences) Scholarship (1995 – 1998). Two extended study periods in Greece were funded by a Greek Government (IKY) Scholarship (1998/99) and an ERASMUS Grant in association with Thessaloniki University (1997). The writing up of this thesis was completed whilst funded by Studentship (1999) of the British School of Archaeology at Athens (BSA). I would like to thank my official supervisors, Professor Paul Hancock and Dr Iain Stewart. Additionally, I would like to acknowledge the advisory role of Dr Graham Shipley who, upon the death of Professor Hancock in December 1998 voluntarily stepped in, reading and commenting on my written work throughout its final stages. I also wish to thank Dr Athanassios Ganas who provided unlimited assistance throughout my thesis research and especially with the logistical hurdles inherent in carrying out fieldwork alone in Greece.

Many other people have assisted in a variety of ways at different stages of my research. In particular I extend my thanks to the director and staff of the British School at Athens; the director, David Blackman, has been both interested and supportive in my endeavours, the office staff provided invaluable assistance in obtaining archaeological permits which would certainly not have been granted without the support of the regional Ephoria (and excavation director at the sites of Kynos and Kyparissi), Dr Phanouria Dakoronia. I also thank the library staff at the BSA for their assistance throughout.

The overriding theme of this thesis has been to present a truly interdisciplinary study and as such requires the use of data and information from fields outside my own particular specialisation. I am grateful to the following people for sharing data with me and/or explaining the nuances of their respected fields (though, of course, I accept full responsibility for the interpretations and conclusions presented in this work):

Dr Athanassios Ganas for remote sensed data (Satellite images) and technical advise with Geographical Information Systems; Dr Angelides who allowed access to and use of the geotechnical report for Atalanti; Dr Stavrakakis, director of the National Observatory in Athens, for supplying the instrumental seismic data; Dr Graham Shipley for advice on Ancient History and translations from ancient/modern Greek and German; Martin Goalen for all things architectural; Professor Nicholas Ambraseys for discussions on the 1894 Atalanti earthquake; Dr Andreas Tertullini for donating copies of the *IGN Seismic Catalogue* (both Italian and English versions) and a dedicated volume of *Annali di Geofisica (Vol X)*; the archivist at the American School of Classics in Athens for access to the Kyparissi (Opus 1911) excavation archive; the archivist at the BSA for access to the Gell notebooks amongst other rare volumes; the British Hydrographic Office in Taunton for supplying the British Admiralty Chart (no 1556) and giving permission to reproduce the extract of it within my thesis; Professor John Coleman (director of CHELP) for permission to use data from the Cornell University excavations at Alai and 'hiring' me for the 1997 excavation season which gave me time to get to know the Lokrian terrain through extensive wandering.

On a non-work related (or personal) level, I would like to acknowledge the encouragement and support provided by my friends throughout my university career. I would like to thank my parents and my brother, Mark, for their continued encouragement and financial support. And finally, I'd like to thank Michael for encouragement and doing all the mundane stuff whilst I prepared the manuscript for publication.

# Abstract

iii

Identifying earthquakes as benchmarkers in archaeological stratigraphies is part of an old tradition of site and excavation interpretation dating back to the early archaeologists of the 19th century. In the past two decades a surge of research in seismology and seismic hazard (both in terms of the resulting effects as well as the earthquake itself) led to increased interest from geologists for the potential of archaeologically derived seismic information to seismic research.

Following this initial burst of enthusiasm it was quickly realised that there were problems with the nature of archaeoseismic data, not least the subjective identification of damage and interpretation of the mechanisms of 'supposed' destructions. The initial response to these problems was to try and construct systematic identification or criteria to provide what could be considered a check list approach to be used by both experienced geologists (who were inexperienced in archaeological aspects) and archaeologists (who often had little knowledge of earthquake geology). However, the similarity between seismic and non-seismic damages made identification of the mechanism of destruction ambiguous, thus plunging the embryonic subdiscipline of archaeoseismology into methodological stagnation.

In light of the above, the subject of this thesis was to evaluate archaeological remains as a proxy data source for earthquake geology and palaeoseismology. This was carried out through an examination of the previous work (both archaeological and geological) to try and identify the underlying causes of the apparent stagnation. Based on the findings of the methodological and theoretical discussion an alternative to the universal identification criteria approach is proposed. This advocates a wholly interdisciplinary approach and is further explored with the application of a Geographical Information System (GIS) as a means to full integration of the disparate data sets, i.e. numerical and textual.

The proposed method is then tested by means of its application to two representative case studies; the inland (settlement) site of Kyparissi and the coastal site of Alai in Atalanti, central mainland Greece, with particular attention to the type of seismic information that can be extracted from the archaeological remains. In the process of testing the working hypothesis a regional database of information is constructed, which although an integral part of the research carried out for this thesis, can stand alone as a reference database.

# Archaeoseismology in Atalanti, Central Greece:

# Table of Contents

# List of Figures

# List of Tables

# Chapter 1: Introduction

The term 'archaeoseismology' is derived from 'archaeoseismic', first introduced as a figure caption in a Nature paper in 1977 (Karcz et al., 1977). The authors, both Israeli geologists, subsequently used the term 'archaeoseismicity' in their critique of archaeologically-derived seismic data in 1978. Although these papers are generally recorded as having defined the term by later workers (for example, Nikonov, 1988 and Stiros, 1996), there is no explicit definition contained within either of them (Karcz & Kafri, 1977; 1978). Not until the publication of the proceedings of an interdisciplinary meeting held in Athens in 1988 was archaeoseismology clearly identified as an emerging sub-discipline of palaeoseismology (Stiros & Jones, 1996). As such it was defined within the foreword of the resulting proceedings volume as follows:

> ... superimposed upon the broad palaeoseismological record, interest focuses on individual seismic events occurring at precise moments over relatively recent time (the last few millennia), whose action affected precise locations - human constructions and their environment - which in turn can be studied in detail through the archaeological record.
>
> (Stiros & Jones, 1996. p1)

A more concise, but essentially similar, definition recently appeared in the textbook titled *Palaeoseismology* by McCalpin (1996) and was based on the author's interpretation of the 1978 paper by Karcz and Kafri:

> Archaeoseismology is the study of prehistoric earthquakes based on their effects on man-made objects, usually buildings or other monuments.
>
> (McCalpin & Nelson, 1996. p81)

McCalpin goes on to qualify the term 'prehistoric', pointing out that the cultures that would be sufficiently advanced to construct lasting structures would also have left written records. This technically advances them into the historical era. This comment touches upon the previously identified and much debated question of where prehistory (or archaeology) stops and history starts. It is worth noting that many archaeological investigations take place on sites whose identity is inferred from the classical literature, such Troy (Schliemann, 1880), and as such often supplement and expand historically or text derived information.

In the above definition by Stiros and Jones (1996) the use of the term 'individual' (referring to seismic events) is somewhat ambiguous. This is because it is not always possible, when dealing with the archaeological record, to gain a resolution that would allow the differentiation of a single earthquake event from sequences, or mainshocks from foreshocks and aftershocks, where these occur. Similarly, 'precise' implies a definitive accuracy in both the chronological dating and the geographical location of

an earthquake. Archaeological dating, although advanced in the use of relative, sequential and scientific techniques (such as the array of radiometric methods), can only in very exceptional circumstances give an exact date in calendar years (Renfrew & Bahn, 1991). For earthquakes, it is more usual to obtain a *terminus ante quem* (date before which) or a *terminus post quem* (date after which) or a calibrated radiometric date range (i.e. date with error margins). Further, it is often a historical source, such as an inscription or literary account that supplies a specific date. For example, a Latin inscription at Corinth (Greece) dates an earthquake that affected the ancient city in the first century to AD 366 (Corinth *VIII, ii, no. 504*, fragment B). The terminology used by McCalpin (1996), when accompanied by the qualifier for 'prehistoric' provides a better overview of what is realistically possible.

In the broader sense the Stiros and Jones definition is preferable to that of McCalpin, because it guides the researcher to consider both human constructions (which by definition include cultural, material, and structural remains) and the environment in which they are located. This contrasts with McCalpin, who rather loosely implies that it is only architectural remains that can provide evidence of earthquakes having affected a particular site. If we accept the overlap between archaeology and history, we can compile the following as a complete definition of the scope and aims of archaeoseismological research;

*Archaeoseismology is the detailed study of pre-instrumental earthquakes that, by affecting locations of human occupation and their environments, have left their mark in the archaeological record.*

## 1.1 Archaeoseismological indicators

In essence, archaeoseismology can be more simply defined as 'earthquake archaeology' or the 'archaeology of earthquakes' (Guidoboni, 1989). That is the use of archaeological investigation to establish details such as location, date and magnitude of pre-instrumental earthquakes. It is the archaeological investigation that, through the presence of observed empirical data, seeks to confirm the seismic event. But, what exactly is the empirical evidence? What do archaeologists and earthquake geologist actually observe which results in their interpretation of earthquake as the mechanism of damage?

As a general background there follows an overview of the type of observations that have been used as evidence for a site having experienced seismic activity. It should be noted that this is not a definitive list but aims to provide the reader with a general overview of the most commonly cited, and often the most visual, evidence used by archaeologists; historians; and earthquake geologists.

Similarly, no critique or order of preference is given for the evidence listed in the following subsections.

FIGURE 1.1 FIELD PHOTOGRAPH OF THE OFFSET TRAVERTINE IRRIGATION CHANNEL AT PAMUKKALE, TURKEY. THE ARROW INDICATES THE SLIP VECTOR OF THE FAULT, DERIVED FROM THE CHANNEL OFFSET (PHOTOGRAPH COURTESY DR R CHALMERS, USED WITH PERMISSION).

### 1.1.1 Architectural Offsets

This heading covers a host of different evidence, from clear breaks in linear anthropogenic remains, either lateral or vertical, for example breaks in the irrigation channels at Heirapolis (Pamukkale), Turkey, see Figure 1.1, (Hancock & Altunel, 1997), to lateral shifts, or misalignment, of architectural elements. Examples of the latter include the shift of upper parts of walls from the base foundations identified in the Late Helladic III period in Kynos, Central Greece (Dakoronia, 1996) and the misalignment of the internal blocks lining Thracian tumuli near Vurbitsa, Bulgaria (Christoskov et al., 1995).

### 1.1.2 Skeletal Evidence

Crushed skeletal remains of humans and animals found in unlikely contexts (for example under piles of rubble or fallen columns) or, specifically in the case of humans remains, in unceremonial or non ritual burial settings. Although perhaps the least common category, examples include a group of three that appear to be huddled and covered by limestone rubble in Kourion, Cyprus (Sorren, 1981) and individuals crushed by the falling columns of the North Stoa of the Agora in Cyrene (Bacchielli, 1995).

### 1.1.3 Evidence from Columns

This group includes a variety of rotational observations resulting from the misalignment of fluting or other decorative features; lateral offsets for example in the Temple of Hephestos in the Agora, Athens (see Figure 1.2); the parallel alignment of a series of fallen columns, for example at the site of Susita, Lake Galilee, (Nur & Ron, 1996); and the unidirectional imbrication of drums

of individual fallen columns such as those observed at Olympia (Dinsmoor, 1985).

FIGURE 1.2. OFFSET OF INDIVIDUAL COLUMN DRUMS AT THE TEMPLE OF HEPHESTOS, IN THE AGORA, ATHENS, GREECE.

### 1.1.4 Architectural Deformations

Architectural deformations is another subheading that covers a great diversity of evidence from architectural elements (walls, floors, staircases, etc) that appear to be bent; bowed; leaning; tilting; out of line; or any combinations the aforementioned, for example bowed and tilted. The stoa at Kyparissi in Lokris, central Greece, is a commonly cited example of a seismically deformed foundation (Stiros, 1988; Stiros & Dakoronia, 1989). One of the finest examples of this type of damage can be seen in the fortification walls at Messina, Peloponesse. The misalignment of the limestone blocks display a sinuous, wave like, pattern along the extent of the walls and are strongly suggestive of earthquake activity.

### 1.1.5 Archaeological Destruction Levels

Stratigraphic layers that contain a high proportion of destroyed ceramics, buildings material (especially roofing terracottas) and/or other cultural remains (usually luxury items that would have been taken with migrants) are often termed destruction levels and earthquake is often the cited cause of such levels. Examples include the chaotic arrangement of the building materials and the decapitated statue observed at Achinos, Central Greece (Papaconstantinou, 1996) and a laterally extensive deposit of broken terracotta at Alope, Central Greece (Dakoronia, *pers comms*).

### *1.1.6 Submerged (flooded) /Emerged (drained) Remains*

Remains that that are interpreted as no longer in their original vertical position are used as indicators of a change in relative sea level or ground water table (for non coastal sites). Examples in this category include the coastal sites of Palaeomagasin on the Atalanti plain, Alai and the inland site of Dion in Macedonia, Northern Greece. Occasionally, the same vestiges have been used to show that movement is not directionally restricted, for example Dvorak and Mastrolorenzo (1991) identify phases of emergence and submergence from the Roman remains of Serapis in the Pozzouli Bay, Southern Italy.

## 1.2 Aims of Research

The aim of this research is to conduct a qualitative assessment of the validity of archaeological evidence as a proxy indicator of seismic activity, with specific reference to normal faulting. This is carried out through:

1) a discussion of the fundamental theoretical issues underlying cross-disciplinary research projects:
2) the proposal and construction of an alternative cross-disciplinary research strategy; and
3) the practical application of this interdisciplinary methodology to a study area with previously cited archaeoseismic evidence, in order to make a value judgement on both the methodology and the use of archaeologically derived seismic data as proxy indicators.

## 1.3 Structure of the Research Volume

In Chapter 2 I present a concise review of previous methodological publications and examples of research that has used, or is using, supposed archaeoseismic observations in a number of different areas of geological research, incluing, sea level studies and neotectionic kinematics. Chapter 3 presents a discussion of the theoretical and methodological issues relating to cross-disciplinary research, with specific reference to the aims of this research and the integration of science and humanities methodologies. In this chapter, methodological stagnation and the previous practice of using universal identification criteria for seismic damage are discussed, and an alternative project-specific method is proposed. This methodology is then applied to an extensional tectonic regime in Atalanti, central mainland Greece, in order to test the working hypothesis. Chapter 4 presents the data (both direct and indirect) that have been collected and the format of this chapter is specifically chosen to highlight to the reader the fact that carrying out a truly interdisciplinary archaeoseismological appraisal of an area requires a large quantity of diverse data; the result of which is that any subsequent manipulation and analysis are difficult and cumbersome. Chapter 5 discusses the application of a Geographical Information System (GIS) as a tool to address the problem of the full integration of numerical and textual data. Additionally, the ability to convert data into useable information is explored in the development of 'earthquake effects models,' which are proposed as an additional tool for archaeoseismological appraisals. In Chapter 6 two site-specific appraisals are carried out using the methodology presented in Chapter 3. The sites were chosen specifically because they had previously been cited as containing archaeological evidence of seismic damage, using the existing identification criteria and methodology; upon the completion of an appraisal using the methodology proposed in this investigation, they would highlight the benefits of the alternative research strategy approach and the qualitative assessment of the use of archaeological remains as proxy seismic indicators. Finally, Chapter 7 presents a synopsis of the conclusions of the research together with an outline of future research in archaeoseismology, namely logic tree formalisation as a move towards an interpretative approach.

## 1.4 Site Nomenclature

The identification and appellation of sites included in the Atalanti case study are not the purposes of this research. In many cases throughout Greece, site names and spelling have changed frequently depending upon the transliteration of the Greek alphabet. In this work I follow the names and spellings assigned by each excavation director (in published accounts). In certain instances where there has been more than one director, for example Alai, which is also is also referred to as Halae (Goldman, 1940) and Halai (Coleman, 1992), I have used the form which is most familiar to me whilst still maintaining the director's spelling in any direct quotations. Additionally, all the modern Greek stress accents have been omitted in place-names, to enhance the readability of the account. This procedure in no way implies that the author agrees with the appellations of sites; I simply seek to refer to the archaeological data in a systematic manner throughout this work.

# Chapter 2: Literature Review

As an interdisciplinary area of research, archaeoseismology (or earthquake archaeology) poses a classic problem when trying to trace the development of what is now a widely practised subdiscipline: namely that, until the relatively recent appearance of dedicated volumes of international journals or book series, previous work is scattered among the specialised, not always readily available, journals of the contributing research disciplines, across many countries, and in many languages. Similarly, given the diversity of research methods used it is impossible to follow a logical chronological order of developmental steps. It is therefore unsurprising that to date no comprehensive and critical account of the development of the discipline has been published; an individual researcher, although well qualified in one area, cannot critically assess the evidence from all contributors. These difficulties, coupled with the huge volume of literature that has come to light during an extended literature search, makes both a definitive account and critical review of archaeoseismological research impractical. Instead, this chapter aims to provide the reader with a general background to the development of archaeoseismological research in the Aegean region through a systematic review of the major issues and what are considered key 'methodological' papers. To a great extent this follows a chronological order, as this is considered to be the optimal way to rationalise the complex development. Naturally, this review primarily focuses on the available literature which is mainly European and American, and the author acknowledges the absence of literature from Eastern Europe, Asia and Oceania due to inaccessibility either through language barriers or unavailable publications.

## 2.1 Chronological Development

### 2.1.1 The Early Scholars

Many archaeological sites throughout the world, but especially in the Mediterranean, Near East, and North Africa, are known from their inclusion in ancient and/or religious texts. In the Aegean region several of these ancient texts include a discussion of the effects of earthquakes upon cities, battles and invasions, for example Homer's *Iliad*, Thucydides's *Histories*, Strabo's *Geography* and the Bible. Personal experience (Evans, 1928; Naumann, 1971) and contemporary reports of seismic activity (i.e. newspapers such as *The Levant and Eastern Herald* and *The Times* (see section 4.4.2), coupled with the classical literary evidence of previous natural hazards in the vicinity of sites, led to the use of earthquake phenomena for the interpretation of otherwise unexplainable destruction layers and occupational hiatuses discovered in archaeological sites. Rarely, if ever, were the relics considered within their physiographic setting (Knossos being a partial exception) in this site-specific, historiographic, approach to the archaeological remains.

*Schliemann (1880, 1884)*
One of the earliest examples of the use of classical literature to first locate a site, and second explain the apparently discontinuous archaeological stratigraphy, can be seen in the work of H. Schliemann in his excavation of Homer's Troy at Hissarlik, western Turkey. In 1880 Schliemann noted in his appraisal of the first year's work at Hissarlik (in 1871) that 'the stones of these house-walls appeared as if they had been separated from one another by a violent earthquake' (Schliemann, 1880. pp. 21). In his later publication the author not only uses earthquakes to explain the apparent destructions (without citing any actual observations) but also makes a direct correlation between his interpretation of the archaeology and a classical literary reference to previous earthquakes in the area. "*Mr Calvert calls my attention to the statement of Pliny, *H. N. II 86*:…which proves that earthquakes occurred here in earlier times" (Schliemann, 1884. p. 26).

Employing earthquakes as the mechanism of destruction in a purely historiographic manner has continued through to the present, using what are becoming increasingly tenuous links with the classical literature. For example, first excavation reports of the site of Alai in central Greece correlate apparent seismic destruction levels (dated through correlation of architectural terracottas with those found at another site of a similar age in close proximity) with the 426 BC earthquake described in accounts by Thucydides (*iii.89.1-5*), Strabo (*i.3.20*) and Diod. Sic. (*xii.59.1-2*) (Goldman, 1940). Such associations are often made in the absence of any geological considerations, for example epicentral distance and site conditions. The same classical references to the 426 BC earthquake have been used to account for an apparent destruction level identified in the Kerameikos and the Agora in Athens (Rotroff and Oakley, 1992), and in seeking both the cause and the date of the destruction in the Kerameikos, the excavators appear to have amalgamated two separate earthquake descriptions described in Thucydides (one in the winter [427/6 BC] and one in the summer [426 BC]) into a single event. This has led the archaeologists to the interpretation that the later 426 BC (or summer event) was the cause of the destruction observed in the Kerameikos. Again, the authors take no account of the geological practicalities of this conclusion, for example the distance of over 100 km between Athens and the assumed epicentral area (the Maliakos Gulf), which, as Ambraseys points out, would make the destructive earthquake effects 'beyond the limits of the possible' (Ambraseys, 1996).

The above discussion is not an argument against an earthquake as the mechanism of destruction of the ceramics in the Kerameikos (I have made no primary investigation or observations at this site), but is included merely to illustrate the particular and persistent problem arising from using text references (classical, historical or contemporary) indiscriminately in multidisciplinary research. When a text is taken out of context, or not used comparatively with other proxy datasets, the results of the interpretation can be misleading and at worst completely inaccurate. Certainly, practitioners from differing disciplines will be unfamiliar with the circumstances in which text information is committed to the record, and with how information can be used by an author in a myriad of ways to support a particular point of view, or belief, while apparently maintaining historical integrity. Similarly, correlation between events of roughly the same date can seem especially plausible when physiographic and geological considerations are not taken into account, especially in terms of text references to natural disasters.[1]

Clearly, with this 'historiographic' approach to seismic destructions in archaeological remains, the archaeologist is rarely concerned with the earthquake itself or its parameters. Instead it is the date (or period) when it occurred that is of interest, and it is often assumed that observed destructions can be corroborated directly from literary accounts or indirectly through damaged cultural remains, i.e. ceramic typologies etc. From a methodological perspective, the archaeologist is seeking to construct a culture or occupation history for a particular site from the accumulation of archaeological finds. Once an earthquake has been identified, through observations and/or literary references, the archaeologist will incorporate this event within the reconstruction or interpretation of the occupation history of the site. The 'dated' natural catastrophe is now part of the overall historiography of the site, and its inclusion in the excavation reports *implies* a factual basis for what is essentially a subjective interpretation of site observations.[2] Problems arise when earthquake geologists then take this 'archaeoseismic' reference as primary empirical data for earthquake catalogues and subsequent seismic-hazard assessment or seismic-risk analysis. As a result, the pre-instrumental earthquake catalogue has a propensity towards inaccuracy through the inclusion of 'fake' or 'rogue' events identified by the over-zealous; in the other extreme, events can be missed due to the over-sceptical views of an excavator.

Despite the broad archaeological tradition of using the occurrence of an earthquake as a *deus ex machina*[3] interpretation of archaeological sites, a number of studies, prior to the development of the 'new' scientific archaeology in the 1960s, had quite unique approaches. Perhaps the best example is the excavation by Sir Arthur

Evans, in the early 1920s, at the site of Knossos, Crete, which has been considered to be the forerunner of present archaeoseismological studies (Stiros, 1996), certainly in Greece. A second example is the work of Claude Schaeffer,[4] who further developed the idea of a seismic chronology based on identified destruction levels within the archaeological site stratigraphy.

*Evans (1926)*
The traditional view of Evans as the founding father of archaeoseismology (Stiros, 1996) is based upon Evans's use of earthquakes destruction horizons as stratigraphic benchmarkers. In turn these markers are then used as key elements in the interpretation of different phases and cult practices observed at Knossos. Evans's own experience of a damaging earthquake (Evans, 1928; [26 June 1926, IX Mercalli Scale] Papazachos & Papazachou, 1997), coupled with extensive research into historical events and geological data led to the production of what, even by today's standards, can be considered as an integrated multidisciplinary report. Perhaps the only missing elements are the inclusion of detailed geological information, such as lithologies, site damage maps and derived earthquake parameters, such as intensity. From the text and footnotes of Evans's monograph *The Palace of Minos, Knossos*, there is no doubt that he conducted extensive research in or was well read in areas such as history, classics, geology, and earthquake effects, including observations of seismic damage in the broader geological context of the island of Crete in his final monograph (see Evans, 1928. pp 312 - 325). Notably, he also comments on the possible use of damaged architectural elements to indicate the direction of propagation of the seismic waves: 'The upper part of a masonry pillar of recent construction which was moved bodily several centimetres due South supplied, indeed, a good index of the prevalent direction from which the waves of disturbance came' (Evans, 1928. p 316). The inclusion of such an observation indicates that Evans was aware of the current thinking in earthquake geology, perhaps unsurprising given the polymathic background of many of the early antiquarians.

As discussed above, it is Evans's use of earthquakes as 'stratigraphic benchmarks' (i.e. destruction horizons datable from the juxtaposed archaeological material) that has been singled out as his major methodological contribution and it is for this reason that he is considered the founding father of modern archaeoseismological studies (Stiros, 1996). However, as discussed above the concept of earthquakes as 'stratigraphic benchmarkers' was already established in the traditional site-specific archaeological approach (see Schliemann, 1884). Instead, it is arguably the inclusion and derivation of geological information from the damaged archaeological remains that can be considered his major innovation, together with the integrated nature of his reports. In short, Evans

---

[1] The interested reader is directed to (Eagleton, 1996) for a clear discussion of literary theory and to Ambraseys and White (1996) for a discussion of the use of historical records and literary accounts in palaeoseismic research.
[2] See van Andel (1994) for a general discussion on the use of scientific data within archaeological reports
[3] This term means a power, event, person or thing that come in the nick of time to solve a difficulty; potential interposition, used especially in literary drama

[4] Translation of the original text (from French to English) courtesy of Armelle Dion. MA.

appears to consider earthquakes as an archaeological interpretative tool whilst maintaining his understanding of the seismic event as a geological phenomenon. This general theme was further developed by C. F Schaeffer in his archaeological researches in the eastern Mediterranean.

*Claude F. Schaeffer (1948)*
The relationship between urban destructions and the sudden development of the bronze industry and its prodigious spread among the Near East and protohistorical Europe at the beginning of the second millennium led Schaeffer to consider the catalytic role that earthquakes may have played. However, as he himself notes in the foreword of his monograph that he was rather reluctant to adopt such a seismic explanation, mainly due to his lack of experience in 'such matters' and the experience of modern earthquakes where disaster is not preceded by the sudden migration of peoples or desertion of settlements (Schaeffer, 1948. p *x*).

In order to redress his apparent lack of geological understanding and to allow a thorough assessment of archaeological research and excavation observations, Schaeffer informs us (1948. p *x*) that he first had to familiarise himself with tectonic geology (as it was understood at the time). From this base, and following archaeological research, he proposed site-specific chronologies based upon identifying earthquakes in the archaeological stratigraphy. In most cases these destructions could be relatively, and, in exceptional cases, absolutely dated on the basis of associations with text accounts or context evidence. For example, the 1365 BC event at ancient Ugarit observed in the archaeological stratigraphy was correlated with a seismic disaster included in a report to the then Pharaoh of Egypt, Amenophis IV.

Again, the importance of Schaeffer's work lies in the direct and explicit application of current geological understanding and the acknowledgement that earthquakes are not site specific and therefore must be observable in other sites of a similar age within a geographical extent. Thus, Schaeffer discusses not only the number of events that had been recorded in this 'Mediterranean geosyncline', but also the integration of geological research/knowledge with the archaeological affects. Additionally, Schaeffer (1948. pp *xiii*) rationalises the decrease in the violence of earthquakes by incorporating the ideas that earthquakes correct the 'imbalance of earthly masses' and that the 'intensity of earthquakes inherently leads to a decrease [of the imbalances] as the equilibrium becomes more stable'. Although the geological ideas put forward in this text appear somewhat fantastical upon first reading, many were concurrent with the geological understanding of the period, that is prior to the concepts of plate tectonics and continental drift. Further, although many of the these ideas are now considered incorrect, this should not detract from the underlying archaeoseismological methodological

contribution of *inter-site correlation* of supposed destruction layers to create a site chronology based on an inter-site seismic stratigraphy. Unfortunately, this idea of cross-site correlation was not actively picked up and has lain dormant for much of the later archaeoseismological work in Greece, which has concentrated on intra-site identification criteria.

Major changes in archaeological theory saw a move from the traditional to the processual approach in the early 1960s and into the post-processual theories of the late 1970s up to the present. Coupled with these major theoretical changes came a drive for archaeological explanations that were explicit and objective. This manifest itself in the introduction of 'borrowed' concepts and techniques especially from the social and natural sciences (Renfrew, 1991). Almost certainly this fundamental shift away from the traditional historiographic accounts of excavations towards a more 'multidisciplinary' approach was a contributing factor towards the use of archaeoseismological evidence in archaeological research. An example of this new cross-disciplinary research can be found in the work of Dales (1966), an archaeologist who presented an alternative hypothesis for the decline of the Indus civilisations based on integrated archaeological and geological data, namely placing sites within their settings

*Dales (1966)*
Dales (1966) proposed that the Harappan civilisation centres of Harappa and Mohenjo-daro did not perish at the hands of Aryan invaders (as previously accepted), but were more probably the victims of the consequences of environmental change, specifically the flooding of the valleys in which the major civilisation centres were located. Dales proposed that the mechanisms behind catastrophic floods were topographic changes resulting from geological disturbances which affected the natural drainage patterns in the Indus Valley. The investigating team included archaeologists, sedimentologists and a hydrologist and combined the sedimentological record with the archaeological evidence to provide what can be perceived to be a more balanced interpretation of the site (Dales, 1966). Despite Dales's work, this type of collaboration was not the norm during the initial period of interdisciplinary archaeology (certainly in Europe), and because of this the development of archaeoseismology as a coherent subdiscipline tended to remain regionally fragmented; this was also a result of the goals of specific research projects.

### 2.1.2 Archaeoseismology and Pre-Instrumental Catalogues

The previous section reviewed the role of earthquakes in the traditional historiographic approach to archaeological interpretations and touched upon the impact of the theoretical changes associated with the 'New Archaeology' of the later 1960's and early 1970s. These fundamental theoretical shifts in archaeology coincided with major advances in geological knowledge, especially in dynamic earth processes (for example, the acceptance of continental drift and recognition that global seismicity patterns correlate with the new global tectonics hypothesis), which arose from major post-war technological advances (specifically in geophysics and sub-surface exploration).

New branches of geological research flourished in the wake of these major advances, not least those which it was hoped would enhance quality of life, such as the eventual prediction of the probability and severity of large magnitude seismic events in order to reduce the cost to society in both human and economic terms (McCalpin, 1996). This was believed possible by gaining a thorough understanding of the generic cause of earthquakes through better documentation of current seismicity patterns (seismology) and extending the window of observation of earthquake activity into the pre-instrumental era (Ambraseys, 1972) for probability of recurrence.

*Ambraseys (1960s)*
The prerequisite of seismic hazard research (into the probability of occurrence and severity of future earthquakes in any given locality) is a long, and ideally continuous, period of observation of seismic activity. For the pre-instrumental era this is usually achieved by the use of macroseismic data, i.e. the written accounts of observed damage effects of earthquakes (Ambraseys, 1975). The Mediterranean region possesses one of the longest and comparatively well recorded histories, and it is this area which has been the focus of a historical seismological research programme led by Ambraseys at Imperial College, London, since the early 1960s. The methodological approach and problems of using macroseismic information in historical seismology are well documented (Ambraseys 1983, Guidoboni *et al.*, 1989; Ambraseys & White, 1996, Guidoboni, 1996,), and there are now many examples of historical catalogues from several countries around the Mediterranean basin, including Italy (Guidoboni and Ferrri, 1989); North Africa and the Middle East (Ambraseys, 1996, Ambraseys & Jackson, 1990, Ambraseys & White, 1996); Greece (Papazachos & Papazachou, 1989; Papazachos & Papzachou, 1997); Israel and adjacent areas (Amiran *et al.*, 1994) and Palestine (Kallner-Amiran, 1951).

As a means to extend the seismic database beyond the historical timeframe, and also to corroborate the very early textual references, (e.g. Biblical and ancient texts) the shift to archaeologically-derived information seemed to be a natural progression (Karcz *et al.*, 1977, Karcz & Kafri, 1978). As with the macroseismic data used in historical seismology, the direct use of archaeologically-derived earthquake information presents a number of similar problems, including subjectivity (especially within interpretation of observations of alleged damage), contextual misinterpretation of archaeological data,

exaggeration in the texts, and positionality of primary data (Ambraseys, 1975. 1983).

The progression into palaeoseismology and the use of preinstrumental data (both historical and archaeological) coincided with the emergence of the 'new' science-based archaeology which seized upon opportunities to employ a barrage of scientific, mathematical, and computer-based techniques in an attempt to gain a better understanding of the archaeological site (Renfrew & Bahn, 1991). However, many archaeologists had little or no experience in many of the new techniques they employed, which in the field of archaeoseismology resulted in a mixture of empirical fact with assumptions validated by appeal to literary extracts (Ambraseys, 1973). Similarly, seismologists and earthquake geologists had little experience in handling anything but numerical data and the results of these initial forays into cross-discipline collaboration (i.e. science with arts) were often badly integrated archaeological monographs or publications with a naïve faith in the data from the unfamiliar discipline. For example, many dates of earthquakes were lifted unquestioningly from texts and excavation reports with little attention paid to the techniques of dating. Nevertheless, there were several collaborations in which natural scientists and archaeologists worked together to produce a coherent discussion of the material and a feasible hypothesis with a seismic element. These can be described as the first interdisciplinary research in archaeoseismology (see section 3.4.5 on the comparison of interdisciplinary and multidisciplinary research).

*Sorren (1980s)*
A notable example of high-profile accessible interdisciplinary archaeological research that uses earthquakes in site interpretation is the work of Sorren at Kourion, Cyprus, in the late 1970s and early 1980s. Based on an uncorroborated hypothesis that the Roman civilisation at Kourion was destroyed by an seismic event sometime in the fourth century (dates of AD 321, 332 and 342 were advanced as the serious contenders), an excavation permit was granted to Sorren for the specific purpose of testing his theory that the destructive earthquake occurred much later, possibly in AD 365. This later date had been derived from the examination of previous finds (numismatic evidence dated no earlier than AD 364, and text evidence both ancient and modern). The finds from the 1984 campaign were beyond those anticipated: extensive destruction was observed in the relatively small excavation area (6m *x* 4 m). Building upon these discoveries (both in the field and the archive) and a correlation between other Cypriot sites and environmental evidence, Sorren put together not only the historiography of the site, but also a general regional picture of seismicity at this period (Sorren, 1981). From the published literature it is easy to identify the reasoned and logical progression of the inquiry; the problem is clearly identified following a background investigation; specialists in separate disciplines contribute to analysis of specific materials, for example an architectural historian,

| Factors for Consideration of Supposed Seismic Damage | Discipline |
|---|---|
| Location and size of site | G/A |
| Main periods of occupancy | A/H |
| Age of Damaged structures | A/H |
| Nature of excavation works (rescue and salvage operation, preliminary, single season, continuing, etc.) | A |
| Mode and mechanism of excavation (equipment employed, amount of overburden removed) | A |
| Extent of excavated area and number and size of exposed buildings and structures | A |
| Type and quality of construction of the damaged buildings and structures (e.g. masonry, stone, adobe, etc.; type of cement, reinforcements and fundaments | AT |
| Type of damage (e.g. collapse, orientated collapse, tilting, breakage, subsidence, fractures, displacement) | G/E |
| Extent and distribution of damage across the site (number of damaged elements, changes in amount and intensity of damage, direction of features of damage and of any possible alignment of the fallen components) | A |
| Occurrence of similar damage to other contemporary sites | A |
| Differences between observed features of damage and those characteristic of man-induced damage | G/E |
| Physiographic setting of the site (relief, distance from cliffs and slopes, slope characteristics, distance from watercourses and shores) | GG |
| Type and composition of the ground (e.g. rock, alluvium, clay; depth to bedrock, etc.) | G |
| Features of recent ground instability (e.g. slides, creep, rockfalls, desiccation cracks, erosion gullies and rills, occurrence of karst features) | G |
| Structural setting of the site (e.g. distance from faults and their orientation, occurrence of joints and their orientation, inclination and structural position of the strata) | G |

Table 2.1: The scheme of description of supposed archaeoseismic damage proposed by Karcz and Kafri (1978), with the column to the right showing the discipline best qualified to do the research, as defined by the present author. The diversity of disciplines listed clearly indicates that archaeoseismological appraisal is inherently a cross-disciplinary research area. **A** = Archaeology, **G** = Geologist, **H** = History, **AT** = Architecture, **E** = Engineering, **GG** = Geography/Geomorphology

an epigrapher and several geologists from the Cyprus Geological Survey. Additionally, this is followed by the clear publication of the work within a reasonable time frame after the initial research.

Additionally, and perhaps uniquely, the material is published at a number of levels, i.e. in the popular media (*Archaeology* and *National Geographic*), in specialised journals for specific areas of research (*Report of the Department of Antiquities of Cyprus*), and as discrete research project in the book *Last Days of Kourion* (Sorren, 1981). Superficially each publication addresses a separate topic and suggests a traditional historiographic approach to archaeological research at Kourion. However, from the diversity of the publications and the fact that all the separate topics are assessed in relation to additional specialist knowledge (for example the crushed

skeletons are analysed in the context of their burial), it is clear that there was an exchange of specialist knowledge between the different workers on the project. The interaction between the separate specialists has allowed Sorren, as project co-ordinator, to interpret the archaeology within the limits of the other proxy data. Further, the geological inferences from the archaeology are set against the geological data provided by the specialist, and a number of senarios are discussed before the preferred hypothesis is outlined. Sorren systematically reviews the merits of the previous destruction dates derived from text, epigraphic and numismatic evidence in the light of dialogue between himself and the epigrapher. Further, in his 1981 publication, Sorren discusses the archaeological data in the context of the information provided by the geologists and geophysicists. Without doubt this research represents a truly interdisciplinary[5] (*sensu* van Andel, 1994 and section 3.4.5), even though it is essentially historiographic in origin.

### 2.1.3 Earthquake Archaeology

As noted above, one of the major goals of palaeoseismologists, or earthquake geologists, is to construct an extensive and reliable catalogue of seismic activity both present and past. In this context archaeological sites were initially viewed as the verifiers of the written accounts which were perceived as being 'plagued by superstition and exaggeration' (Karcz *et al.* 1977). However, the difficulties of correlation between archaeological field observations and literary evidence were quickly recognised by geologists (Ambraseys, 1971; 1973), and in 1978 Karcz & Kafri published the first critical appraisal of archaeological earthquake evidence.

### Karcz & Kafri (1978)

In this work the authors assessed the validity of attributing a seismic mechanism in archaeological remains identified by observations such as tilting, collapse, sagging, subsidence, fracturing and fissuring in walls, buildings, structures and pavements. This was carried out through a discussion of alternative destructive processes, such as adverse geotechnical effects (i.e. differential subsidence and mass movements for example landslip or rock falls), poor construction conditions and human influence. They concluded that 'In spite of their significance and usefulness, the archaeological data cannot be employed as an entirely independent technique for the verification of ancient chronicles and the study of past seismicity' (Karcz *et al.*, 1977). The reasons are clearly stated as, operator's bias (i.e. subjectivity), bias in historical information, and more importantly the contribution of geotechnical (site) effects in the natural course of decay of ancient structures.

A systematic method of description of damage was advocated as a means to maintain 'the proper balance between the geological, geomorphological and geotechnical factors, i.e. the natural environment, and the historic and anthropogenic considerations' (Karcz & Kafri, 1978. p237) that were noted as factors affecting the identification of earthquakes as the mechanism of destruction: these are summerised and presented in Table 2.1. The additional shaded column highlights the cross-disciplinary nature of archaeoseismology by noting which discipline would be best qualified to carry out the research However, it should be noted that although Karcz and Kafri identified the problems associated with identifying seismic damage in archaeological remains, they did not propose a specific methodology which would assist the archaeologist or geologist in the integration and ultimate interpretation of the contrasting datasets resulting from such a proposed scheme of description (for

example the chronological history and the neotectonic setting of the site).

### Rapp (1986)

As noted at the beginning of this review, earthquakes were identified as a destructive force in the ancient city of Troy as early as the 1880s. In the late 1970s, a review and reassessment of the evidence for, and nature of, seismic activity of the Troad was carried out by George (Rip) Rapp. The results of this review were published as a separate supplementary monograph (No. 4) of the initial excavation reports published by the University of Cincinnati (edited by Carl Blegen) from the beginning of the 1950s. In order to provide context to the findings of the review of the evidence for earthquakes at the site of Troy, Rapp included sections on '*Quantifying Earthquakes and Effects*' and '*Seismicity in the Region of the Troad*' (Rapp & Gifford 1982). The authors acknowledge the work of Karcz and Kafri (1978) on the difficulty of identifying and assessing earthquake damage in archaeological sites, and propose that much of the ambiguity could be resolved through analogy of the effects of modern large-magnitude earthquakes upon structures similar to the ancient ones. A review of the 'Archaeological Evidence for Earthquakes' is followed by an 'Assessment of Probable Earthquake Activity at Troy' in which Rapp gives an analytical framework (shown in table 2.2) within which to interpret structural damage in archaeological remains. The framework essentially is a distillation of the natural surface processes outlined by Karcz and Kafri (1978), with no consideration of the human factors (or cultural transformation processes) that could be in operation, either past or present.

**Analytical Framework Proposed by Rapp (1982) for the Interpretation of Structural Damage in Archaeological Remains**

*Assessment of:*
1. the mechanical properties of the building materials;
2. the nature and quality of construction;
3. special characteristics of the regolith (overburden), including topography, earth, soil materials, and hydrology;
4. the earthquake regime; and
5. archaeological evidence of destructive human forces.

**Table 2.2:** The framework within which to interpret structural damage in archaeological remains proposed by Rapp (1982).

In a subsequent paper, Rapp (1986) developed the theme of integrated collaborative archaeoseismological work. He acknowledged that mechanisms of destruction can also have an anthropogenic origin and noted that 'unravelling events recorded in archaeological remains has been hindered by interpretations **not based on any rigorous criteria**' [my emphasis]. In his conclusion,

[5] See Chapter 3, section 3.4.5 for a discussion of interdisciplinary research.

9

## I) Man-induced Action
### A) Destructive Ones
    1   warfare
    2   fires
### B) Those due to Construction
    1   partial dismantling or complete pulling-down of a structure
    2   restructuring, reconstruction, repair

## II) Natural Processes
### A) Slow Destructive Processes
    1   weathering, washout, cracking and other surface effects
    2   landslide, subsidence, gravitational warping, tilting, deformation
    3   shrinkage, heaving and other deformations in structure-bearing rock and ground
    4   washout and destruction by shore, bank, and slope processes
    5   slow tectonic movements
### B) Rapid (Catastrophic) Processes
    1   rockslides, landslides, snow avalanche
    2   showers of rain, mud flow, floods, tsunami, storms
    3   hurricanes, whirlwinds, air storms
    4   earthquakes and associated deformations

## III) Historical/Archaeological Considerations
### A) Periods of Economic Prosperity and Stagnation
    1   increased construction & commercial activity, culture expansion
    2   decreased activity, population emigration, contraction of inhabitation areas
### B) Periods of Anthropogenic Destructions
    1   warfare (including methods, fortifications, religious buildings)
    2   destruction of buildings (settlement specific/regional)
### C) Periods of political unrest/change of rulers
    1   ethnic changes or population migrations
    2   change in religious and industrial units from numismatic and cult evidence
### D) Construction technology
    1   methods, materials, design
    2   architectural procedures and standards

## IV) IV Data Collection
### A) Quality of data
    1   primary/secondary or other
    2   contemporary with event
    3   degree of reliability

**Table 2.3:** The scheme for consecutive examination of the causes of damage (collapse) in architectural and archaeological monuments. From Nikonov (1988).

Rapp again recommended comparative studies of modern earthquake damage on architecturally similar structures; detailed geological and physiographic maps; and a set of diagnostic criteria to confirm the earthquake as the mechanism of destruction. Perhaps the most important message to be read from this work, which has recently been re-emphasised by workers such as Ambraseys (1996; 1996b), Di Vita (1996), Guidoboni (1995; 1996), and Noller and Lightfoot (1997), is the need for a 'multiple working hypotheses approach' (Rapp, 1986. p 375) or simply, to be aware of alternative reasoning. It is this final point which is carried further by Nikonov (1988).

*Nikonov (1988)*
In this paper, Nikonov addressed the three separate but closely linked issues concerning archaeoseismological

research: the potential, the principles, and the techniques and associated problems of archaeoseismic research. He also noted the justification of archaeologically-derived earthquake data; (1) in terms of monument protection and conservation, (2) as a means to validate written accounts and therefore extend historical seismicity, and (3) for use in earthquake-hazard research in terms of providing more realistic evaluations of recurrence intervals and seismic risk.

Essentially, Nikonov's paper presented an expanded discussion of the scheme proposed by Karcz and Kafri in 1978, but emphasised the human contributions to destruction, not simply in terms of war and fire, but also in terms of economic declines, pestilence, etc. Additionally, he provided an extremely comprehensive list of specific processes within the section dealing with natural mechanisms, including numerous factors divided into slow and rapid processes. Table 2.3 shows the scheme, as proposed by Nikonov (1988), which should be adopted in the examination of architectural and archaeological monuments.

Nikonov's scheme is accompanied by a detailed and greatly expanded discussion of each element, specifically the human factors (cultural transformation processes) which were mentioned but were not extensively considered by Karcz and Kafri (1978), who were instead biased towards an engineering geology reassessment of previously published evidence. In contrast, Nikonov suggested that graphs of settlement development, including for example periods of war, economic prosperity and political unrest, should be plotted and used as reference datasets against which the dates of any proposed earthquake damage could be considered. It is clear from these proposals that Nikonov was an advocate of a wholly interdisciplinary approach, and his scheme suggests that from the outset of a project each set of possible damage mechanisms should be fully investigated by a specialist in that particular field of research. For example, the occupation and economic history should be carried out by a historian and/or an archaeologist, and the surface processes by an engineering geologist. Arguably, the main criticism of Nikonov's scheme for the description and documentation of archaeoseismic damage is one that the author himself points to in the conclusions: namely, that the sheer number of factors to be considered and investigated for each possible occurrence of seismic damage makes the presented scheme highly labour-intensive and practically unworkable. No abridged 'user-friendly' methodology has since been proposed, which implies that the loss of any particular factor in the investigation would reduce the thoroughness of the investigation and compromise the results of the findings. As Nikonov (1988. pp 1319 - 1320) notes, 'The above considerations and concepts [outlined in the paper] are viewed by the author as a basis, a version to be discussed and improved. A consistent and usable version can only arise as a result of team work'.

### 2.1.4 Archaeoseismology in Greece

During the early stage of the development of archaeoseismological research, all the methodological papers (Karcz & Kafri, 1978; Rapp, 1986; and Nikonov, 1988) either explicitly state or implicitly suggest the positive advantage of a regional analysis of proposed seismic damage through interdisciplinary research. However, having ascertained the range of research required, they all fell short of suggesting a feasible or workable methodology that would achieve an effective integrated study of the obviously disparate datasets. What is apparent from the available literature is that subsequent developments in archaeoseismology vary considerably from country to country, and even between separate research groups.

The simplistic response to this would be that the degree of development reflects the availability of previous literature (i.e. principal methodology papers) and the strength of previously established interdisciplinary contacts. In Italy, for example, historical seismological research carried out shows strong collaboration between seismologists and historians, thus the move into the prehistoric period has built on existing collaborations. (See the dedicated volume *Earthquakes in the Past* [1995] in the journal series Annali di Geofisica which focuses on multidisciplinary research and in several cases achieves interdisciplinary results, for example Christoskov *et al.*, (1995) and Gergova *et al.* (1995). In Greece multidisciplinary links were not quite as well established and only limited collaboration can be seen more usually in the fields of coastal and sea-level research. (Stiros & Dakoronia, 1989; Pirazzoli *et al.*, 1994; Stiros *et al.*, 1996). However, it should be noted that there is a general reversal in this trend, in keeping with the recent increase in cross-discipline research projects, for example the extended Cornell Halai and East Lokris Project (CHELP) in central mainland Greece and the joint Environmental Project in the Corinth area by Ohio State and Vanderbilt Universities,[6] which are considering the archaeological sites within both the present and the palaeo-environment. Similarly there has recently been a steady increase in the number of palaeo-environmental studies in several areas in central Greece and the Peloponnesus by workers such as van Andel (1989), and Zangger (1991) which also consider the archaeological data.

The gradual introduction of archaeological observations as indicators of recent tectonic movements on terrestrial structures is well illustrated in the work of Stiros in the Atalanti area of central mainland Greece. In a preliminary report (Stiros, 1985), all the known archaeological sites on the Lokris (Atalanti) coast were surveyed as a means of deriving the relative change in sea level associated with recent tectonic movement on the Atalanti (Lokris) fault. Although previous neotectonic research by Lemeille (1977), Rondogianni-Tsiambaou (1984) and Rondoyanni (1988) had mentioned the presence of submerged archaeological sites in the region, Stiros (1985) provided the first survey that attempted to quantify the amount of sea-level change using archaeological relics. Each known coastal archaeological site is listed with a brief discussion of the date of the physical remains and an (estimated) quantitative measurement of the relative change in sea level. Stiros concluded that the archaeological remains 'definitely supports the [geomorphological] evidence of significant relative sea level change' (Stiros, 1985. p 18), but also noted that limitations in the data, i.e. knowledge of the original height above sea level of the vestiges, lead to a

---

[6] Both these projects can be viewed on the WorldWideWeb at the following addresses:
CHELP at http://halai.fac.cornell.edu/ and KRRC at http://eleftheria/stcloud.msus.edu/krrc

poor constraint upon the amplitude of the vertical motion and consequently only qualitative observations.

Subsequent research led to the inclusion of observations of earthquake damage in terrestrial archaeological sites in close proximity to the active fault, as a means of collating the coastal indicators with the ancient literary references to earthquakes that affected the area (Stiros, 1988a; Stiros, 1988b; Stiros, 1988c; Stiros & Dakoronia, 1989). Further, attempts were made to derive the mean slip rate on the fault calculated using both the observed earthquake damage and classical literary references (Stiros & Rondogianni, 1985), and although the figure has now been revised by some recent geological studies (e.g Ganas, 1997), the use of archaeologically derived earthquake data had been firmly established in neotectonics research in Greece.

Thus, although initially used as indicators of relative sea-level change (both tectonic and eustatic components), by the mid 1980s archaeological observations were also being used as indicators of surface faulting, inferred from dislocations of relics (Stiros, 1988a;1988b). However, the identification of seismic damage in archaeological sites, such as structural offsets or deformations, was still in the domain of the archaeologists, albeit with the observations being imported into neotectonic research.

It is argued in this research that inexperience in both geological sciences and engineering seismology, coupled with inherent theoretical incompatibilities of the methodologies used to obtain the diverse data-sets, are the biggest factors contributing to limited application of archaeologically derived earthquake data. In past archaeoseismological appraisal many archaeologists relied on their excavation training and personal experience to guide their judgement about the observed damage that may have had an earthquake mechanism. However, a serious problem with such subjective observational judgements is the possibility that earthquakes could become a *deus ex machina* explanation of what previously would have been unexplainable damages or destructions. It is interesting to note, that although archaeologists were identifying earthquakes in their excavations, there do not appear to have been any suggestions from the archaeological community with regard to the establishment of a methodology for either identifying or carrying out a thorough archaeo-seismological appraisal.

*Stiros (1980s)*
In an attempt to bridge the disciplinary divide in technical understanding between archaeology and geology, Stiros (1988a) presented an overview of geological aspects of earthquakes and surface faulting in relation to archaeology. Although not dissimilar to the paper by Rapp (1986), who reviewed the geotechnical elements that should be considered in archaeoseismological research, Stiros provided an earthquake geology base with an overview of several case studies from central

Greece (primarily the Atalanti (Lokris) region) in order to demonstrate the feasibility of identifying fault movement/geometry from damaged archaeological relics. This general paper was presented at an archaeological conference in Athens (*New Aspects of Archaeological Science in Greece*), and appears to have been specifically for the benefit of archaeologists. However, in the simplicity of his approach, Stiros fails to mention the complexities of geotechnical effects (also known as 'site effects') upon constructions and the paper generally implies a straightforward site-specific observational approach to the identification of proposed seismic damage. Significantly, there is no reference to placing specific site observations within the broader context of the regional tectonic regime or geomorphological and geotechnical factors; considerations which had been outlined by both Kafri *et al* (1978), Rapp (1986) and Nikonov (1988), as essential to a full archaeoseismological appraisal. The site-specific (damage) approach was further developed by Stiros and Dakoronia (1989) with the publication of some observational diagnostic criteria and this was later refined and updated by Stiros (1996) (see table 2.4), who presented diagnostic criteria for the recognition of ancient earthquakes in archaeological remains and accompanying proviso that the reader take into account the reservations expressed in other sections of the text. Although primary effects (such as surface faulting) and secondary effects produced by shaking are mentioned, the main thrust of the reservations stated focuses on the responses of the actual structural remains to earthquakes (with modern analogies) and implies a more engineering-based seismological approach.

Although structural response is an important aspect of cultural resource management, preservation and conservation, it is important to note that the vast majority of archaeological remains are not complete structures. More usually they comprise configurations of single features (walls) dispersed architectural elements and artefacts (for example pottery sherds) which may also have experienced a number of external stresses (not only of seismic origin) and may well have been the subject of previous repair and strengthening which would influence their response to any current seismic activity.

In summary, the literature pertaining specifically to Greece shows a trend towards the compilation and application of diagnostic criteria as a means of reducing subjective ambiguity in identifying observed seismic damage. Although it is necessary to identify damage within a site, this specifically intra-site (individual site only) approach cannot be accepted as a universal methodology for archaeoseismology; it fails to consider external factors, for example human actions and influences (outlined by Nikonov (1988)) and the contribution of natural factors, for example foundation material as discussed by Rapp (1986; 1988). Additionally, this identification criteria method can only be used in areas where there are architectural relics,

**Criteria for the identification of earthquakes**
**from archaeological data**
(in keeping with the reservations expressed in other sections of the text)

1. Ancient constructions offset by seismic surface faults.
2. Skeletons of people killed and buried under the debris of fallen buildings.
3. Certain abrupt geomorphological changes occasionally associated with destructions and/or abandonment of building and sites.
4. Characteristic structural damage and failure of constructions
5. Displaced drums of dry masonry columns
6. Opened vertical joints and horizontally slid parts of walls in dry masonry walls
7. Diagonal cracks in rigid walls
8. Triangular missing parts in corners of masonry buildings
9. Cracks at the base or top of masonry columns and piers
10. Inclined or subvertical cracks in the upper parts of rigid arches, vaults and domes, or their partial collapse along these cracks
11. Downslid keystones in dry masonry arches and vaults
12. Several parallel fallen columns
13. Several fallen columns with their drums in domino-style (slices of salami) arrangement
14. Constructions deformed as by horizontal forces (rectangles transformed to parallelograms)
15. Destruction and quick reconstruction of sites, with the introduction of what can be regarded as 'anti-seismic' building construction techniques, but with no change in their overall cultural character
16. Well-dated destructions of buildings correlating with historical (including epigraphic) evidence of earthquakes
17. Damage or destruction of isolated buildings or whole sites, for which an earthquake appears to be the only reasonable explanation.

**Table 2.4:** Criteria for the identification of seismic damage in archaeological remains. From Stiros (1996)

as the main thrust of the identification focuses on architectural features, for example offset structures or deformed walls.

*Guidoboni (1989)*
Although not specifically a methodological advance, the publication of the first earthquake catalogue (Guidoboni, 1989) combining both historical literary and archaeological data (which provides an accessible reference volume for pre-instrumental seismic research in the Mediterranean region) can be seen as a major step in the development of archaeoseismological research. The result of many years' work by a fully interdisciplinary team, it provides a clear and critical evaluated list of all the known pre-instrumental earthquakes. The clarity of presentation of the information, including small location maps (showing the probable extent of damages as inferred from the literary accounts), full citations (with an indication of the reliability of the references) and a commentary on the sources, makes this the most comprehensive catalogue published to date. Although initially published in Italian, it was subsequently translated into English (Guidoboni 1989) (albeit without the collection of contributory papers by historians and archaeologists from around the world i.e. France (Helly), Greece (Stiros & Dakoronia) and Crete (Sorren)), making it also the most accessible volume to date. Without doubt, the availability of a standardised catalogue of textual references should be considered a valuable step towards a standardising the methodology of archaeoseismology. However, as with all chronological catalogues, its reliability will only be maintained through a system of updates as new research brings to light more information

and necessitates amendments. Ten years after its publication, no such amendments have yet been published.

**2.2 Current Trends**

As noted previously, the purely historiographic descriptive approach is still very much in use with examples found throughout the recent literature. In the recent *Archaeoseismology* volume (Stiros & Jones (eds) 1996) approximately 70 per cent of the contributions essentially deal with individual sites at an interpretative level. Additionally a large percentage of the work focuses on Aegean sites (more specifically Greece), which probably reflects the classical-archaeological origins of the research and the relative youth of the discipline. However, the bias may be simply due to the fact that the conference was the result of collaboration between a member of the Institute of Geological and Mineral Exploration (IGME) and the then director of the Fitch Laboratory at the British School at Athens, and was hosted in Athens. In contrast 90 per cent of the papers given at the recent international conference '*Volcanoes, Earthquakes and Archaeology*', held in London (1997), displayed a more multidisciplinary, though not necessarily interdisciplinary approach (*sensu* van Andel, (1994) (see section 3.4.5 for a discussion of the terms interdisciplinary and multidisciplinary). The reasons behind these differing approaches are numerous, not least that archaeology has traditionally been 'topic-focused' in terms of specific sites, regions, or finds, for example Mycenae, Peloponnesus, ceramics or mosaics. Other specific congress or conference volumes, for example *Earthquakes and Building Cultures in History: New Disciplines and Applications: Ten Years of Research* (Rome 1993) exhibit a similar trait, i.e. that of archaeoseismological research moving towards a multidisciplinary ethos.

The lack of a central database, and the wide distribution of the material in terms of publications, make it impossible to review here all the cases of archaeological research that are using, or have used, supposed earthquake damage or destruction as a means of interpreting an archaeological site or as part of wider seismic research. Recent examples include work on the construction of a site history of ancient Cyrenaica, based on the archaeological context of crushed skeletons and dated using numismatic evidence from coin hoards (Bacchielli, 1995), the change in power relations in Minoan Crete following seismic effects (LaRosa, 1995), and the work of Klaus Kilian correlating settlement reorganisation and changes in building and pottery styles with the occurrence of seismic events (Kilian, 1996). Much of this type of information is buried in archives, and systematic searching of the literature or personal communication from participating archaeologists are the only means of discovery. Increased awareness of such sites through the increase in the number of dedicated volumes, conferences and workshops and collaborative studies with geologists will slowly bring them to the attention of those working in neotectonics.

Guidoboni (1996) and Di Vita (1995) both noted the apparent lack of dialogue between researchers in different fields, and recently attempted to set up a nucleus of specialists to promote collaborative research in the fields of historical and archaeological seismology (primarily in Italy). One of the major goals suggested by the group was to establish a reference databank of observed seismic effects in archaeological remains for use in multidisciplinary research. Guidonboni, (1995) proposed a 'thematic atlas' of seismic collapse in archaeology as a means of reducing the subjectivity in identification of seismic damage by comparison with other sites with seismic collapse, thus contributing towards the protection and conservation of the archaeological heritage. However, neither the databank or thematic atlas has progressed further than the proposal stage at present, primarily due to the apparent lack of interest and co-operation from the archaeological community in Italy (Di Vita, 1995).

However, without an intensive exchange of ideas from the planning stage through publication and beyond, collaboration between differing disciplines will lack mutual understanding and an appreciation of specific methodologies and perhaps more importantly of the inherent limitations of the resulting data. As van Andel (1994) notes the products of uneasy coalitions between science and archaeology are often incomplete and rarely address the initial questions for which the research was initially instigated.

### 2.2.1 Breadth of study

This penultimate section aims to draw the reader's attention to the breadth of research projects that using, or have in the past employed, archaeologically derived proxy data, either in a fully integrated cross-disciplinary study or simply as corroborative or supporting evidence. This does not claim to be an exhaustive account, as the literature sample reflects to a great extent the aims of the research carried out, and is, as previously highlighted, geographically selective towards the eastern Mediterranean.

*Fault Indicators*

The practice of using architectural remains as indicators, and more importantly quantifiers, of fault movement was reasonably well established in the seismological community before the appearance of the methodological papers discussed in the previous sections. For example in Ghana, West Africa, an archaeologist was called in to date a petroglyph[7] (approx. 3,000 years BP [1950]) on the Accra fault plane which was found located 32ft below the elevation of the normal river surface of the Volta Dam (Thompson, 1970). A vertical displacement of 32ft within the last 3,000 years was derived from the dating, of the archaeological evidence. Petroglyphs are still used as a valuable source of dating and have recently been used by (Karakhanian *et al.*, 1998) to date fault displacements and volcanic eruptions in the Khanarassar fault zone, Armenia, and Hancock and Barka (1987) use a mid-Roman low relief carving in brecciated marbles on the Yokusyol fault plane in western Turkey, to derive an estimate of neotectonic movement.

Similarly in Russia (Trifonov, 1978), the sense of motion and amount of movement on faults in the Kopet-Dagh fault zone, central Asia, were determined from the offsets in the walls of a fortress dating from the Middle Ages and fifth century irrigation channels. Other recent examples of earthquake geologists using this offset relics/remains application as motion indicators can be found in China (Hou *et al.*, 1998), the Kyrgyz Republic (Korjenkov, 1997), Israel (Marco, 1997), Turkey (Hancock & Altunel, 1997), and Greece (Stiros 1988a; 1988b).

*Kinematics*

The continued use of offset remains or archaeological piercing features[8] as an investigative technique by earthquake geologists is mainly due to the misconception that a relatively small amount of general archaeological and historical background knowledge is required. It is assumed that any necessary information, such as dates of construction phases can be gleaned from historical and

---

[7] Rock carving
[8] Archaeological piercing features are discrete elements (structural or cultural) within an archaeological site that are offset or dislocated.

archaeological literature, or if necessary requested from a specialist in the required subject once the field data has been collected. Guidoboni (1996) cautions against using specialists from other disciplines at the conclusion or interpretative phases of a project, as the newcomers would be unaware of important background information such as data collection, storage, etc. However, geological investigations that unexpectedly reveal useful archaeological finds, such as petroglyphs on fault planes, preclude initial collaboration at the planning stage and calling in a specialist after the find is better than not using one at all.

Special care should be taken during the interpretative stage of archaeologically obtained data as the most immediately apparent observations may not necessarily lead to the correct hypothesis. For example, an initial examination of the offsets in the travertine irrigation channels at Pamukkale, Turkey, inferred strike slip[9] faulting with sinistral motion. However, this does not fit with general range-front tectonic setting of the region, and upon reappraisal of slip vectors (derived from the offsets in the archaeological relics) it was found that the sense of motion on the faults was clearly normal[10] with a very minor element of oblique movement (Hancock, *pers. comm.*). Clearly disparate data sets need to be assessed in conjunction with each other, and where anomalies occur between a well established hypothesis and the results of a new line of enquiry, e.g. using the archaeological remains as a proxy indicator of neotectonic activity, a careful and integrated reassessment needs to be made.

In contrast, good interdisciplinary research projects can reap detailed results. For example, an interdisciplinary team of archaeologists and geologists working in North America, where written history is relatively recent in comparison with the old world (i.e. in the last 1,000 years), applied an archaeoseismic approach to the derivation of earthquake parameters on strike-slip faults. In the published reports the authors first note that a fundamental element of archaeoseismic appraisal is a clear understanding of the variety of responses of archaeological remains (as cultural deposits or structural relics), depending upon the influencing tectonic structure: for example, lateral offset associated with strike slip, and both horizontal and vertical movements associated with the diverse geometric variations that can be observed with both normal and thrust faults. This approach is exceptional at present, in that it seeks to derive offset data from purely subsurface cultural imprints of human occupation and tectonic structures with purely lateral movement. The truly interdisciplinary nature of this project (*sensu* van Andel, (1994) is clear, in that the project gives equal weighting to the geological and archaeological data, considering each as complementary data sets providing information missing from the other.

Noller and Lightfoot (1997) use cultural piercing features, or points of known dislocation, to derive the cumulative amount and mean slip rates along strike slip faults, with the emphasis placed on the dating benefits of cultural archaeological material over geological, 'non dateable' piercing features, such as stream channel offsets. Additionally, in certain circumstances they suggest that it may even be possible to define a 'per event' slip by observing individual offsets in specific occupation layers (archaeological stratigraphic horizons). It should be highlighted that this research is primarily concerned with using 'offsets' (termed cultural piercing features) in occupation layers. Notably the identification criteria would be totally inapplicable in this instance, due to the complete lack of surface structural remains and the lack of monumental architecture or distinct architectural remains in Native American settlement sites. This highlights the point made earlier that identification criteria are specifically applicable to structural remains (primarily found in old world archaeology)[11] and cannot be assumed to be a universal method of archaeoseismic appraisal. Noller is currently developing his archaeoseismological research in a project started in 1998, which applies a similar truly interdisciplinary methodology to the extensional regime of central Greece, specifically in the region of the Gulf of Corinth (Noller, *pers. comm.*)

A similar methodology has been applied to the Dead Sea transform fault zone using piercing points in the architectural remains of the Ateret Crusader Fortress (Marco & Agnon, 1995; Marco, 1997). The potential of using offset architectural remains in other tectonic environments, such as the extensional tectonic setting of central Greece, has been previously discussed and is the focus of several publications by Stiros (1988). However, it should be noted that archaeological remains straddling a well defined active fault are a rare occurrence in Greece, almost certainly due to the well established differential topography of the opposing fault blocks (i.e. hangingwall and footwall). In environments where there is established seismogenic topography it is more usual to find the archaeological remains situated on either the lowered fault block (hanging wall), in the case of sanctuaries, or in the raised fault block (footwall), in the case of fortifications or acropolis. Settlement sites often take advantage of the pronounced topography, with the fortifications located in the high ground and the occupation site on or near to the accompanying plains for example, ancient Corinth, and ancient Athens. Exceptions to this occur where there is little or no pronounced surface manifestation of the active structure, i.e. there is no visible displacement in the ground cover. One of the few examples of vestiges traversing a possibly active (but

---

[9] Faults with slip vectors parallel to their strike... the term lateral is a synonym for strike slip and is often used. The descriptors 'sinistral' and 'dextral' refer to the sense of motion, left and right respectively.

[10] Normal faults are identified by the downward movement of the hangingwall in respect to the footwall.

[11] The Mediterranean, Asia Minor etc., as opposed the New World archaeology of the Americas.

as yet uncorroborated) fault was the temple of Poseidon in the Corinth plain (Noller *pers comms*).

As noted above, faulted relics, a term synonymous with offset remains and cultural piercing points, have also been used by Hancock and Altunel (1997) to redefine the geometry of an active segment of a normal fault within the Hierapolis fault zone, western Turkey. The major archaeoseismological implications of this site-specific study were that, although offset structural remains were useful markers for the determination of deformation patterns, faulted relics alone are less reliable for dating episodes of slip and therefore quantifiable slip rates. Therefore, this study corroborated the work by Noller and Lightfoot (1997) who suggested that it is the sediments that contain the dateable 'cultural' remains which are useful in determining the amount of mean slip on a specific point, and not the physical architectural remains themselves. However, this not only requires the site to straddle an active fault, but also requires co-operation between archaeologists and geologists at the initiation and planning stage of the research. It is doubtful that such good datasets could have been achieved if collaboration had been established in the post-excavation stage of the investigation. Guidoboni, (1996) discusses in detail the need for collaboration from the earliest stage of excavations and points out that without this the identification of seismic damage in archaeological sites will rarely go beyond conjecture, due to an inability of the specialist (geologist, geophysicist) to evaluate the methodological correctness of the hypothesis. This point is explored further in the following chapter on theory and methodology.

*Coastal Research*
Although the potential of using archaeological remains as indicators of relative changes in land and sea level was not truly realised until the late 1960s, observations of submerged (and emerged) remains had been made by the early antiquarians, for example Negris (1904), and provided the background to the later work of coastal geologists such as Flemming (1968; 1969; 1973; 1986), and Pirazzoli, (1992; 1994; 1996). Research carried out by Flemming from the late 1960s sought to define eustatic (global sea level) changes in sea level in the Mediterranean using archaeological remains (Flemming 1969). However, extensive field survey in coastal and submerged sites throughout the Mediterranean basin highlighted the role of tectonics in the observed relative change in sea level (RCSL). In the majority of cases of submerged, or emerged, sites it was noted that the eustatic[12] factor was negligible in relation to the tectonic influence. Further, it was found that the amount of vertical displacement was random, depending upon the geographical locality of the site in relation to the active structures (Flemming, 1968; 1969; Flemming *et al.*, 1973; Flemming & Webb, 1986).

This type of coastal research is entirely dependent on the use of archaeological remains as a vertical datum of palaeo-sea level from which proxy measurements of vertical motion can be derived. There are, however, problems with using coastal relics, and Blackman (1973) pointed out the need for careful scrutiny and cautious interpretation of the archaeological remains, whose identification and original position on the coast may still be a point of debate. For example, stone quarries and causeways may equally have been located in the tidal zone or partly underwater. Additionally, Blackman (1973) notes that some features, i.e. stairs, tanks, tombs, cisterns, etc., can rarely if ever quantify the absolute change in sea level, due to the imprecise location on the coastline, whereas others such as quays, mooring stones, moles, etc., can define only a maximum and/or minimum change. Further, Blackman (1973) urges interdisciplinary work as the way forward, but with a brief cautionary note of the limitations of these data; again, no specific methodology was proposed. Flemming (1973) similarly distinguishes two main groups of coastal archaeological features. The first are those that must have been located on or near to the shoreline and which must have remained above or below the sea. The second group belong to structures that depend on the tidal fluctuations and marine conditions for their use (Pirazzoli, 1996. p 53). A key factor in coastal studies that incorporate archaeological derived data is the presence and role of errors, especially accumulated errors.

As with any research, the results of sea-level studies which use archaeologically derived data may be negated by the presence of cumulative errors relating to every aspect of the observations, i.e. the tidal range, the position of the feature on the coast and the tectonic contribution of vertical movement. Errors associated with these and other factors can often be greater than the resolution of the indicators themselves. For example, stone quarries located within the intertidal zone in the Atalanti bay on the southern margin of the Gulf of Evvia may have subsided by approx. 10 cm in the last known seismic event on the Atalanti fault (April 1894). Given that the tidal range is between 0.25 and 1.0 metres along that section of the coast (derived from coastal observations), it is not possible to accurately distinguish the neotectonic element of relative sea level change simply because the tidal range is greater than the movement of the indicators. Similarly, the use of regional sea-level change curves such as those published by Flemming and Webb (1986) and Lambeck (1996) and can produce highly misleading results in localised site specific sea-level studies, due to the coarse resolution of the regional data imposed upon a relatively small study area (Rapp & Kraft 1994). Ideally, research involving sea-level indicators should seek to verify or calibrate the regional curves with local data prior to making any quantitative calculations.

*Archaeological Interpretation*
A thought-provoking overview and discussion of the use of natural disasters in relation to archaeological

---

[12] Eustatic change in sea level refers the relative change associated with non tectonic factors, i.e. the glacial melt contribution

historiographic interpretations, and indeed to the construction Mediterranean prehistory, is given by Drews (1993). Although the primary objective of the book is to discuss 'New Warfare' as the primary mechanism for the end of the Mediterranean Bronze Age, he discusses in detail previous hypotheses, including a chapter dedicated to the role of natural disasters in the decline of the Bronze Age. The reasoned argument against the earthquake hypothesis leaves the reader with no doubt that the identification of earthquake effects in archaeological remains has been in the past, and is at present, extremely subjective; further, that published hypotheses pertaining to destructions and damage in archaeological sites frequently change as new theories on past human behaviour are introduced. Perhaps one of the key points made is the distinction between 'damage' and 'destruction' by natural phenomena and the likely human behavioural response to this, i.e. remain and repair, or flee and start afresh. The probability of earthquakes affecting sites throughout Greece, specifically in the southern Peloponnese, is not contested by Drews (based on the fact that the area is in a zone of high seismicity). However, the hypothesis of large universal events, or 'killer quakes' (Drews, 1993. p 43) which are often used by archaeologists to explain any laterally extensive damage and the decline of whole civilisations, e.g. the Aegean Bronze Age (Mycenaeans/Minoans) is dismissed on the grounds of a lack of supporting archaeological evidence and inconsistencies between datasets from the individual sites. Clearly, Drews's discussion is not unbiased in the presentation of the data, because he is writing to support his hypothesis of new warfare techniques as the reason behind the decline of the Mycenaeans. However, his chapter raises some interesting points that caution against the unquestioning inclusion of archaeologically identified earthquake events in any further seismic studies or earthquake catalogues.

## 2.3 Discussion

In a broad context, it would appear that, once the problems of using archaeologically derived earthquake data had been established in the early 1980s, methodological advancements fell into a period of stagnation. During this period publications often re-emphasised the need for a standardised methodology through the identification of additional reasons or factors that contribute to operator bias (Guidoboni, 1996). Similarly, the need for a set of observation or identification criteria was also reiterated in work that attempted to incorporate structural (i.e. buildings and construction) engineering with the existing corpus of material (Stiros 1988a; 1988b; 1993; 1996).

During the last two decades the number of volumes dedicated to archaeoseismology has increased, but the majority of papers describe the identification of a seismic event in an archaeological site, rather than presenting new methodological developments in archaeoseismological research. As noted above, in Greece archaeo-seismological research has focused on identifying pre-instrumental earthquakes through the use of observational criteria. This multidisciplinary approach involves the integration of results or conclusions from differing areas of research and only partially fulfils the approach set out by Karcz and Kafri (1978), Rapp (1986) and Nikonov (1988). One of the main problems with establishing a truly interdisciplinary methodology is finding an approach which will allow the full integration and analysis of differing datasets at a variety of scales.

The simplistic view would be that the fragmented development of archaeoseismological appraisal is largely a function of the poor availability of previous literature, the lack of co-operation between disciplines, a lack of understanding of the limitations of the each data set and the lack of a tool with which to integrate and analysis multiple datasets. This highlights a major problem with trying to construct a 'universal' methodology, in that the nature of the investigation is to a certain extent site-specific as a result of a high number of contributing variables, not least the topographic expression of the tectonic setting of the sites under investigation. For example, Hancock (1996) has noted that ancient column drums can be used as analogues of the cylindrical rock samples used in fracture-mechanics experiments, and can therefore be indicators of seismic shaking. In this context, a macro-survey of these relics may act as preliminary indicator of a site 'worthy of detailed investigation'. However, construction materials apart, the geological and geotechnical factors alone at each site will produce very different column responses to an earthquake of the same magnitude with an epicentre the same distance away. Additionally, sites beyond the limits of the classical world, or earlier than the use of monumental architecture, and therefore without the presence of columns, could not be identified using this observation alone. Similarly, sites located away from the coast cannot be identified through the presence of submerged or emerged remains. It is clear, therefore, that the number of regional and site specific variables precludes the development of universal criteria for identifying of archaeoseismic damage, and a universal methodology for identifying seismic damage within the site.

Moreover this 'universal' approach only partially addresses the problems identified by Karcz and Kafri (1978), Rapp (1986), and Nikonov (1988), in that it fails to take into consideration the fundamental methodological issues concerning ambiguity due to subjectivity which are inherent in a discipline that involves an element of interpretation of human activity and human self-expression in terms of cultural and non-cultural remains by individuals. This aspect is discussed further in chapter 3, where the problems of variables negating a set of 'universal' identification criteria are highlighted.

# Chapter 3: Theory & Methodology

This chapter presents a discussion on why it is not possible to construct a universal methodology for cross-discipline research and proposes that by using a project construction template a tailor made research strategy can be developed to address the initial questions posed. The theory has been specifically intertwined with the practical construction of the strategy for this project in order to highlight the importance of a dialogue between all the primary considerations in strategy construction.

A research project consists of three repeatable stages which are theory, methodology and interpretation. The flow chart at Figure 3.1 demonstrates that the process of research is indeed cyclical and this is also true in every stage of a particular project (represented by the small grey input/output arrows at the top and bottom of the pages and at the beginning and end of each key stages indicated by the column to the right of the diagram). Moving on to the next stage is entirely the decision of the researcher. This chapter essentially deals with the constituent parts of the first section, the theoretical considerations, although it draws from the results of practical research in the form of previous research

Whilst this chapter cannot be seen as a complete discussion of the philosophical or theoretical issues behind cross-discipline research, it aims to demonstrate that the key to this approach lies in the careful construction of a research strategy resulting in a good empirical database that can be used to construct rational arguments or interpretations by researchers from any of the contributing disciplines. The following chapter four implements the methodology (or data collection) section resulting in the foundation empirical database. This is followed by chapter five which demonstrates the use of the database through specific case studies relating to the refined question defined in the summary of this chapter.

## 3.1 Methodology/Strategy

Before any further discussion it is necessary to outline the conceptual differences in research strategy and methodology discussed within this chapter. A *research strategy* can be described as the overall plan of research including the ultimate aims and goals. A methodology is a 'body of methods or procedures' (O'Hear, 1989) used to fulfil the aims by carrying out the plan or research strategy. Formulating the question, theoretical considerations (arising from a review of the previous research), feasibility, and data identification form the body of research strategy planning; fieldwork and data collection form the main body of methods used in this project. These subsections, together with the final

interpretative section, are shown schematically in the flow chart at Figure 3.1.

## 3.2 Research Strategy

The construction of the research strategy can be broadly divided into three discrete sections. The first deals with theoretical issues and should be carried out prior to the second, methodological, section to ensure that the data collection remains focused on the specific issues of the research. The methodology should be research specific to ensure that the data necessary to answer the aims of the project are collected. The aim of this research was ultimately to make a value judgement regarding archaeologically derived seismic data and the goal of the methodology section was to collect a diversity of data to form an empirical database from which a comparative analysis could be made. The final stage is interpretation which in itself is divided into three levels from low level generalisations of the observed data to the final interpretation, or as mentioned earlier, in the case of this research, a value judgement on the use of archaeological sites as indicators of seismic activity. The remainder of this chapter discusses sequentially the three stages of strategy construction with particular reference to the research carried out.

### 3.2.1 Initial Question

The first step in formulating a research strategy is essentially defining the *Initial Question* and this may or may not be in a traditional question format, for example What, Why, How, etc. The initial question should be, in the first instance, general, as it is the process of refinement through subsequent questions and choices that results in clarity of the overall aims or goals of the research project, i.e. the strategy. Additionally, the research planner should be clear as to the ultimate use of the data, as a lack of clarity at this key stage will inform any subsequent reasoning necessary to plan and carry out the methodology.

The following example demonstrates the importance of clarity. The initial question: *what information can archaeological remains provide about prehistoric earthquakes?* could be posed by two different 'end users'. End users no. 1 (EU1) are archaeologists who wish to carry out a historiographic interpretation of a particular site. In this case the emphasis is placed on:

- Identification of the event at the site (with little consideration of the source of the earthquake, and
- Date, (usually relative within an archaeological stratigraphy).

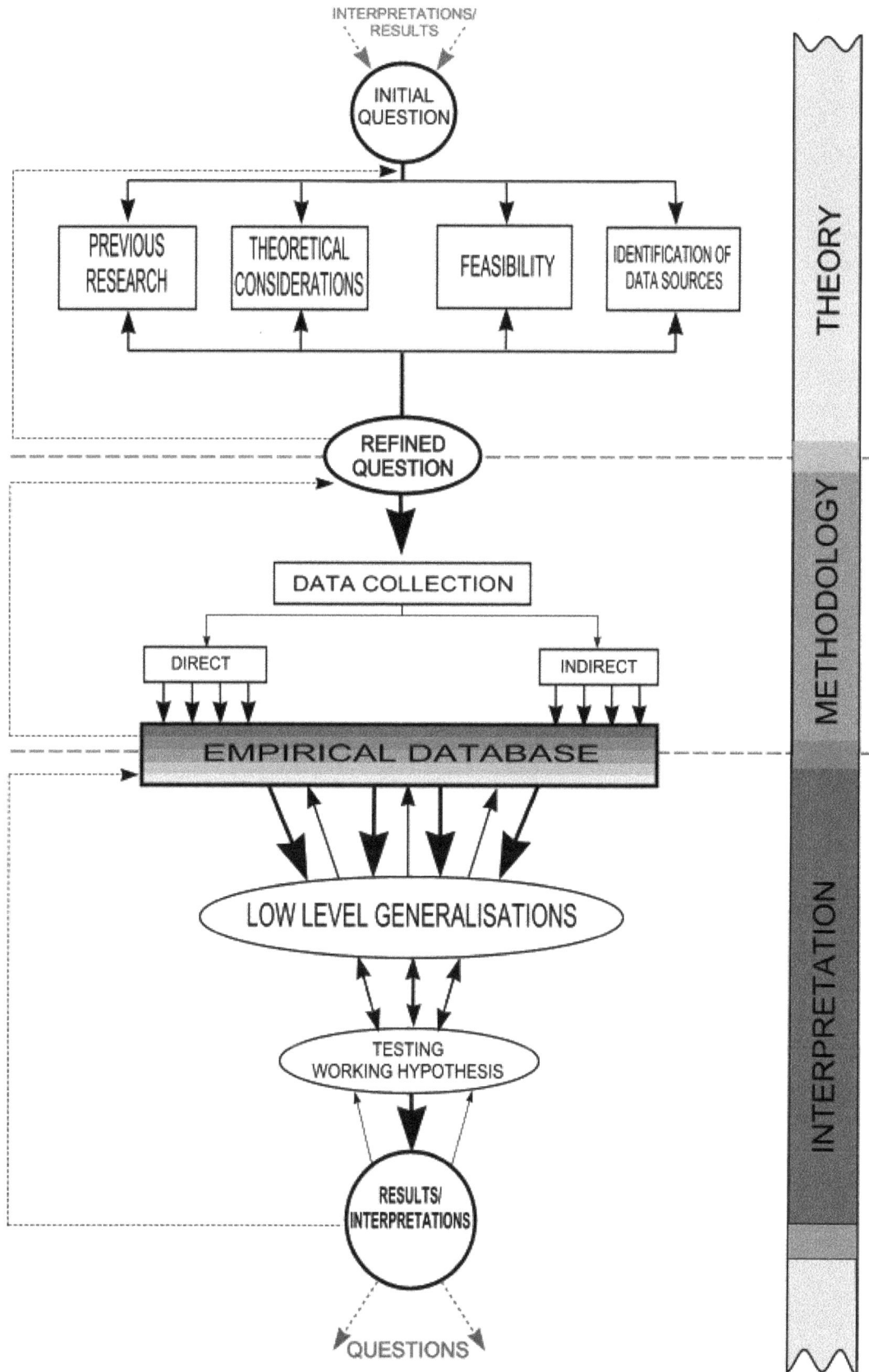

**FIGURE 3.1:** FLOW DIAGRAM REPRESENTING THE PROCESS OF RESEARCH STRATEGY CONSTRUCTION USED IN THIS THESIS. NOTE THERE ARE THREE STAGES, EACH OF WHICH CAN BE REPEATED IF NECESSARY TO ENSURE THE STRATEGY BECOMES WELL INTEGRATED AND WELL FOCUSED

End Users no.2 (EU2) are geologists who wish to use the results of the research to construct a seismic catalogue as a primary data source for use in future research i.e. regional seismic hazard assessments. Their emphasis would be on:

- Size/magnitude
- Time, and
- Location and extent of damage, from which to estimate the magnitude.

As both end users appear to require similar results it could be assumed that the application of a single 'universal' research strategy would result in a methodology that would provide results useful to both parties. However, careful consideration highlights that the archaeological researcher (EU1), requiring only the presence and a relative date of a seismic event within the site, could clearly use data obtained from a geologically-orientated research strategy. However, the geological investigation, (EU2) requiring the derivation of geological parameters, could be rendered inaccurate or misleading by using data obtained from the purely archaeological-orientated research strategy.

Examples of this particular problem are seen throughout the literature, such as the inclusion of rogue, fake or multiple seismic events, in pre-instrumental seismic catalogues, based upon data unquestionably lifted from archaeologically or historically orientated research. Clearly, a combined 'universal' methodology which aims to 'cover all eventualities' would result in superfluous or redundant data sets which were badly integrated in publications. Again evidence of this can be found throughout the literature, for example Goodchild (1968). It is therefore necessary to consider carefully the ultimate use of the research findings as this will allow for the construction of the most effective research strategy and subsequent methodology. The template proposed in Figure 3.1 is designed to guide the planning of a good research strategy through the consideration of fundamental factors at every stage of development.

The aim of this research project was identified as *evaluating the use of archaeologically derived information in palaeoseismological research*. The ultimate end users in this case are earthquake geologists, therefore emphasis is placed on the requirements of geologically orientated research in deciding the essential data layers to include in the empirical data base. However, this does not exclude the use of this information by archaeologists.

As archaeological remains, like geomorphological landforms, have experienced the effects of earthquakes in the distant and near past, it is necessary to define which period [of time] is of interest. Theoretically, although archaeological remains could in effect provide information from the point of their original deposition/construction to the very recent past (i.e. the

previous hour, minute), in reality the maximum benefit of archaeologically derived earthquake data will be to provide information for the period of time when there are few other proxy data sources, i.e. the pre-instrumental *and/or* prehistorical period. It should be noted that making this distinction does not preclude archaeologically derived information as proxy data for other research projects, for example historical seismology. Having defined both the end user and the aims of the research, this can be translated to the following initial question from which to construct a more refined strategy:

*What information can archaeological remains provide about prehistoric earthquakes?*

### 3.3. Previous Research

Apart from the obvious rise in the number of studies using or obtaining archaeoseismologically derived earthquake information, the main theme to arise from the review and analysis is that:

- Methodological stagnation followed the identification of subjectivity within archaeologically derived earthquake information.

Throughout the initial background research it was assumed that the reason behind this stagnation, even with the existence of subsequent methodological papers directing the researcher to multidisciplinary endeavours, was due to:

- The fragmented nature of literature, i.e. research appearing in a wide diversity of publications, and the lack of availability of previous research due to obscure sources or language barriers; leading to,
- A lack of communication between fellow workers in the international community.

However, it has subsequently become apparent that this is a simplistic and superficial explanation of a more fundamentally complex issue concerning the incompatibility of the theoretical backgrounds of archaeology and palaeoseismology. Additionally, it was concluded that methodological advancement cannot hope to be achieved without consideration and incorporation of these fundamental issues in any future archaeoseismological research strategies.

#### 3.3.1 Methodological Stagnation

Although subjectivity within archaeo-seismologically derived earthquake information was identified from an early stage (Ambraseys, 1973, Karcz & Kafri, 1978) it has never been explicitly defined. Initially, the cause of the subjectivity was diagnosed by the natural scientists as the result of the following factors:

- Archaeologists being unfamiliar or unaware of the effects of earthquakes upon both the physical environment and structural/cultural members of the site.

- Archaeologists interpreting the observations in light of previous experience rather than following a specific 'scientific' methodology.

- Archaeologists using earthquakes as a *deus ex machina*[13] interpretation of any unusual and unaccounted observations of destruction.

(Karcz & Kafri, 1978)

The solution to the problem was assumed to lie in the imposition of a more rigorous 'scientific' approach which would result in more empirical, and therefore more objective (scientific), data that could be used in further research. Such an approach would incorporate scientifically obtained observations, such as the responses of certain structures to previous earthquakes or derived from mathematical and computer simulated structural response modelling.

Since the late 1970s the main thrust of the attempts to address the problem has been the development of descriptions of, or criteria for, the identification of supposed seismic damage in archaeological sites (Karcz & Kafri, 1978; Rapp, 1986; Nikonov, 1988; Stiros & Dakoronia, 1989; Stiros, 1996). In each case, multidisciplinary research was advocated as a means to achieve the integration of scientific data, thereby reducing the perceived subjectivity.

However, it is argued here that it is not the archaeologists' apparent lack of understanding of earth science concepts (or vice versa) that is at the root of the of subjectivity (although it is agreed that this would undoubtedly be a contributing factor). Instead, it is the more fundamental theoretical issues and resulting practices that are the major contributors to the stagnation in methodological development.

## 3.4 Theoretical Considerations

Theoretical considerations are project specific and can only be securely identified as a result of background research. From the main theme discussed in chapter two, four areas for discussion have been identified. These are (1) the presence of subjectivity, and (2) how to reduce it, (3) communication and (4) the nature of cross-disciplinary research collaboration.

### 3.4.1 Subjectivity

Objects exist apart from human knowledge or perception of them, they are external to the mind; real and value free.

In archaeology the objects constitute the material traces of human activities which are excavated and processed according to a rational method (Shanks & Tilley, 1987). However, the appellation (of the object(s)) is the result of social creations that allow the researcher (or typologist) to categorise, analyse and subsequently hypothesise about, or interpret, the objects. For example a clay pot is a ceramic vessel; if it is dropped it will undoubtedly break. Based upon arbitrary and selective attributes (often aesthetic qualities, such as colour, shape or form, and pattern) the pot is categorised by the typologists and assumes a separate identity accordingly (for example Halafian, Roman, and BandKeramik). The process of categorisation requires human endeavour and/or human expressiveness (O'Hear, 1989) and as such is becomes a social construct (Shanks & Tilley, 1987)

To be objective is to deal with outward things or exhibited facts uncoloured by feelings, opinions or human expressiveness. Methods (and methodologies) are operations on objects, or the artefacts, and form the basis for explanations, interpretation, theories or hypotheses (Shanks & Tilley, 1987). They are inherently subjective as they require human input to explain the object. Ultimately all explanations or interpretations can be deconstructed, through *reduction*, to the empirical evidence, i.e. back to the object. This is highlighted in the interpretative section of the flow diagram at Figure 3.1, where the evolution or development of a theory/hypotheses can be observed as the product of positive and negative (*bi-directional*) feedback between the objects and the socially created theory. Trigger (1989) describes this as logical coherence and factual response.

If it is accepted that objectivity is the attainment of timeless knowledge (Rowlands, 1982) i.e. without social bias, it follows then that subjectivity is not impartial, but partial, as the explanation or interpretation proceeds from, or belongs to, the individual consciousness or perception; which is in turn a state of being that is a construction of all past and present experiences (Collingwood, in Trigger, 1989). Similarly, Binford, (1982. p 126), describes cognition as 'not direct nor objective, but may be indirect and subjective relative to our beliefs about the world, i.e. our paradigm'. It is not the observations, or objects themselves that are subjective, but the *interpretation* of the empirical evidence, or raw data, which is based upon sum of the interpreter's present and past experience and knowledge. Therefore, in archaeoseismology, as with any other research it is *not* what we observe at a particular site but *how* we interpret the objects that we observe that is at the root of subjectivity. Although not explicitly stated, it is precisely this point that is implicit in the drive towards the construction of methods of description or criteria for identification as a means to reduce subjectivity.

### 3.4.2 Increasing Objectivity

The introduction of methods of description (Karcz & Kafri, 1978; Nikonov, 1988), or criteria for identification

---

[13] See footnote 3 in Chapter 1

(Stiros & Dakoronia, 1989; Stiros, 1996) would, it was proposed by the natural scientists, introduce a standardised, and therefore, objective methodology into the previously subjective identification procedure.

The description proposed by Karcz and Kafri (1978) (see Table 2.1) focuses heavily on the engineering attributes of the site and the environment and physical factors that can be identified, using a universal list, and quantified in perceptible, for example numerical, values for categorising or mapping observations. However, this description considers mainly the impartial factors, and in doing so evades the root of the problem, namely the subjectivity within interpretation of the cultural factors.

This lack of consideration of the cultural factors was addressed in the description for the identification of seismic damage proposed by Nikonov (1988) (see Table 2.3). However, it is immediately apparent that in trying to cover every possible aspect, from the naturally occurring physical properties of the foundation material to the socio-economic evolution of a specific community, he renders the proposed methodology vast, cumbersome, and difficult to use. Additionally, although he draws the reader's attention to the range of factors to be considered in an archaeoseismological appraisal, no suggestions are made for reducing the interpretative subjectivity. However, Nikonov himself identified these problems and qualified his proposal with the reservation that it was a base from which to work.

The most recent attempt at devising a workable methodology is Stiros (1996) and is the only one to explicitly state that it is a 'criteria for the identification of earthquakes from archaeological data' (Stiros, 1996. p 152). Although the criteria sought to provide workable guidance for the field operative, the proposal is based on a typology of seismic effects, again, with a more architectural engineering focus and including factors for consideration that are based on modelling and modern analogies with recent large magnitude seismic events in the Mediterranean region. The apparent simplicity of the scheme is lost with the addition of caveats and the complete lack of attention to the cultural context of the remains. Additionally this criterion misguides the reader into assuming that identification of earthquake damage is a simple case of a going through a universal 'check list' of observations. The inadequacy of this system is highlighted by the following simple comparative study of Doric temples in Greece.

Moreover, whilst it is clear that this 'check list' approach has limited application (it may help assist in the identification of certain types of damage), it cannot be heralded as a *universal method*. Indeed, it is interesting to note that these particular criteria could not be used on the vast majority of sites that do not contain architectural elements, and are therefore perhaps best suited to later Mediterranean cultures which contain exceptional amounts of reasonably well preserved monumental architecture, i.e. temples, baths, stadia, etc. Whereas the criteria addresses the identification of structural damage, it has been demonstrated that the lack of cultural considerations again does not address the human factors and thus the interpretative subjectivity.

### 3.4.3 Comparative Temple Study

In an attempt to assess the validity of such a disassociated approach, often carried out by non-geological/ engineering specialists, a small temple study was instigated. The results show that even with a reduced number of variables to be assessed (in terms of variety of sites and structures) it is not possible to apply a universal criteria for identification based purely on a check list of observations of characteristic damage isolated from environment and the culture historical context f the monuments or relics. Additionally, what may appear to be secure and absolute indicators of seismic damage (fallen imbricate column drums in the case of temples) can conceivably be the result of some other agency of destruction.

*Methodology*

The basis of the study was to demonstrate the subjectivity associated with identifying seismic damage using a 'characteristic damage' checklist approach. To this end it was necessary to isolate the maximum number of variables possible by selecting subjects (archaeological remains) with the maximum number of similar characteristics that had some form of damage which had previously, or could conceivably be, assigned a seismic origin. Once selected, a hypothetical check list criteria (based on that proposed by Stiros (1996)) was carried out in the context normally used by non-specialist field workers, i.e. the structural architectural remains only. Having established the mechanism based on the checklist, a detailed investigation of previously published work for each site was carried out and any observations by archaeologists or historians on the origin of the damage noted. Finally, any recent revisions to the published interpretations (based on recent research) were noted and discussed.

As noted above, in order to assess only the validity of the method (i.e. the check list criteria approach) it is important to reduce the number of variables within the assessment, therefore subjects were considered only if they fulfilled *ALL* the following criteria:

➢ Doric peripteral (6 x 13 columns) temples,
➢ Of a similar age, i.e. constructed between the 4th and 5th C BC,
➢ Constructed of similar material,
➢ Close to seismogenic source (s), which
➢ Contain damage previously cited as of a seismic origin

*Data and Discussion*

The following section is divided into subsections that discuss the data for each subject structure within the study. The data are presented in a standard format and a discussion of the pertinent points follows. The results of the investigations are presented in table forma in the results section of the study.

## Temple One: Temple of Zeus at Nemea, Peloponnese

**Setting:** Situated in an upland valley (approx. 333m a.m.s.l.), in the foothills of the Arkadian mountains, on quaternary unconsolidated sediments.
**Constructed:** @330 BC
**Style:** Doric peristyle peripteral (6 x 13 columns), Corinthian interior colonnade, with Ionic upper order.
**Materials:** Limestones
**Mechanism of Destruction:** Earthquake
**Based on:** The 'salami'-like arrangement of the column drums, radiating out from the krepidoma
**Mechanism of Destruction using a hypothetical identification criteria approach (based on Stiros 1996):** Earthquake
**Discussion:** Descriptions of the Temple by Pausanias (2nd C AD) indicate that although still standing, the roof was missing. This view is supported by the presence of surface weathering on the stylobate. The director of the excavations at Nemea suggests that earthquakes were only a contributory factor after the structure had been weakened by human intervention (Miller, 1990). The absence of the krepidoma (almost completely) supports the human agency of destruction, as this could only be the result of 'robbing out' of the blocks and the metal clamps used in the construction of the columns. Further, evidence of cutting of the lower drums in preparation of 'felling' have been found both on the fallen columns and those still standing. Additionally, the columns are now considered to have fallen at different times, rather than in a single event (Miller 1990, p 131).

**Current View of the Mechanism of Destruction:** *Human Actions*, with possible component of seismic destruction on columns that had previously been weakened by systematic destruction.

## Temple Two: Temple of Poseidon at Cape Sounion, Attica

**Setting:** Constructed on a rocky and steep promontory at Cape Sounion, approximately 60m a.m.s.l.. The Temple is founded partially on thinly bedded limestone dipping steeply to the north, and an artificially constructed terrace.
**Constructed:** @444 - 440 BC
**Style:** Doric peripteral (6 x 13 columns)
**Materials:** Marble and Poros (limestone) foundation
**Attributed Mechanism of Destruction:** Not explicitly stated, amongst those cited, seismic mechanism, strong winds, storms.
**Based on:** Misaligned (displaced) columns drums on the south facade shown.
**Mechanism of Destruction using a hypothetical identification criteria approach (based on Stiros 1996):** Earthquake
**Discussion:** This Temple has over time been reconstructed. In 1676 only 9 columns on the southern side; in 1744 4 remained standing, in 1763 this was reduced to 3, and by 1825 only 2 remained standing. Papadakis (1972) notes that the desolation of the monument 'was not caused by only the passing of the centuries and the tempests; it was mainly the work of man'. This was supported by the evidence of scattered and ransacked architectural remains form all parts of the temple, i.e. the architrave and base.

**Current View of the Mechanism of Destruction:** *Human Intervention*. However, the exact cause of the destruction is not known and it is suggested that the Temple was weakened damaged caused by 'robbing' and the quest for treasure associated with local legend that the site was once a 'Palace of a Princess' which resulted in serious undermining of the foundations (Meletiz & Papadakis 1972).

## Temple Three: Temple of Zeus at Olympia, Peloponnese

**Setting:** The Temple is constructed entirely on unconsolidated quaternary alluvial deposits located within the wide alluvial plain, between the Kronos Hill and the confluence of the Rivers Alpheios and Kladeos.
**Constructed:** 479 BC
**Style:** Doric peripteral (6 x 13 columns)
**Materials:** Fossiliferous limestone, with marble dust stucco
**Attributed Mechanism of Destruction: Earthquakes**
**Based on:** The orientation of imbricated column drums.
**Mechanism of Destruction using a hypothetical identification criteria approach (based on Stiros 1996):** Earthquake
**Discussion:** The Temple is known to have burnt down in AD 426 under the orders of Theodosius II and is quoted as having been destroyed in AD 522 and AD 551 by earthquakes (Photinos 1982). The earthquake hypothesis is further supported by the presence of the krepidoma precluding the demolition of the columns by undercutting the base.

However, Cooper (1992, p 121) notes that 'the same action [vandalism and 'robbing out' of metalwork and blocks] was wrought on the Temples of Zeus at Olympia and Nemea sometime between the 5th and late 6th centuries after Christ'. Given the apparent lack of physical evidence for human destruction a research team from the National Technical University in Athens carried out numerical modelling to ascertain if it was possible for earthquakes to fell the Temple, bearing in mind that this was one of the largest Greek temples in continental Greece. The results indicated that the structural system could 'withstand very strong seismic motion' (Ψυχαρης *et al* 1996 p *xiv*) and the authors conclude that 'in light of these analyses, it appears that unusually strong seismic motions are required to overturn the entire or part of the Temple'. It should be noted however, that the ground, or site effects were not taken into consideration and that the locality of the Temple in the alluvial plain may have been a contributing factor.

**FIGURE 3.2:** MISALIGNED COLUMN DRUMS ON THE SOUTH FAÇADE OF THE TEMPLE OF POSEIDON AT CAPE SOUNION, ATTICA. PHOTOGRAPH BY PEDRO BARROS, USED WITH PERMISSION.

**Current View of the Mechanism of Destruction:** *Seismic and/or Human*

**Temple Four: Temple of Apollo Epikourios at Bassae, Peloponnese**
**Setting:** The Temple is situated on a bedrock of ridge Mount Kotilion at a height over 1000m a.m.s.l.
Constructed: 5th C BC
**Style:** A typical Doric (6 x 15 columns). Peculiarities include: north - south orientation, spur walls set at the diagonal, and a single Corinthian column set in an Ionic colonnade
**Materials:** Finely layered limestone severely fractured with schist like texture.
**Attributed Mechanism of Destruction:** Earthquake
**Based on:** Deformations in the structural alignments and the occurrence of large magnitude earthquakes in the area (Cooper 1965; 1966; 1985).

**Mechanism of Destruction using a hypothetical identification criteria approach (based on Stiros 1996):** Earthquake

**Discussion:** Within this short comparative study the structural remains (both in terms of the archaeology and the preservation) of this temple have been the most extensively investigated. Using the description provided by Pausaius Cooper (1996) notes that the temple was probably in a good state of preservation up until the 2nd century AD, almost certainly with its roof still intact.. Subsequent destruction was almost certainly from 'robbing out' the metals used as clamps, as every block was defaced and the krepidoma and orthostate was also severely damaged and the whole structure weakened. Cooper (1996) places the damage and looting between the 5th and late 6th C AD during the turmoil that accompanied the Slavic Invasions (@ AD 550 - 630). Cooper (1996) notes that any earthquake damage of 'detectable magnitude' occurred during the 1300 year period when the temple slipped into obscurity until re-discovery by the French in 1765 and not as previously assumed, by the 1965 and 1966 events. In 1765 the interior walls were described as being in a ruinous state, though the colonnade was still standing, all except two columns on the south side (a total of 36 columns). Cooper (1996) notes that the only discernible earthquake damage is the lateral offset of column drums, which is constrained to between the second and third drums.

The two large magnitude earthquakes in 1965 and 1966 with epicentres of less than 3km and 17km respectively, caused either little or no effect (Ambraseys 1985, in Cooper 1996) to the temple, or 'provoked serious damage to the temple' (Koukis & Tzitiras 1980, in Cooper 1996) depending which publication is read. Therefore, the present condition of the monument, with regard to the stability and arrangement of the architectural members, was reached by 1927, and Cooper (1996) clearly states that in his opinion the damaging earthquakes occurred during the mediaeval period when the extent of the structural response could not be assessed due to a lack of knowledge of the temple by contemporary scholars.
The occurrence of a double seismic event ($M_s$ 6.0 on 13 September, and $M_s$ 5.4 on 15 Sept 1986 (Papazachos & Papazachos, 1997)) with the epicentres in Eliohorio, east of Kalamata, produced a seismic zone with a radius of 70km, within which Bassai is located. At the Temple site the guard house (60 m from the temple) experienced strong vibrations, whereas the workmen on the Temple itself were 'totally unaware of the event' (Cooper 1996, p 128).

Additional conservation work has highlighted that the major deterioration and deformation of the temple structure is the result of numerous factors, but primarily due to the extreme weather conditions experienced at the site. These include extreme temperature fluctuations that produce mechanically disintegration of the rock;

| Temple Locality | Mechanism cited in original work | Mechanism derived using identification criteria | Revised primary mechanism cited after additional work | Refs cited for revised mechanism |
|---|---|---|---|---|
| Nemea | Earthquake | Seismic | Human Intervention | (Cooper et al. 1983, Miller 1990) |
| Sounion | Natural Phenomena | Seismic | Human Intervention & natural Phenomena | Papadakis, 1972. |
| Olympia | Earthquake | Seismic | Human Intervention | (Photinos 1982) |
| Bassai | Earthquake | Seismic | Human Intervention | (Cooper 1978, Cooper 1996a, Cooper et al. 1983, Cooper 1996b) |

**Table 3.1**: Summary table of the information available for the four temples within the study demonstrating that simple observational identification does not provide an accurate assessment of the mechanisms of destruction or damage in archaeological sites. The criteria used were based on Stiros (1996) listed in Table 2.4.

rainwater causing chemical dissolution and disintegration of the structural elements; and human intervention weakening the structure.

The results of this extensive interdisciplinary research has led Cooper (1996) to conclude that that even in its weakened state the Temple is in more danger from the *non seismic* mechanisms of destruction. He concludes that the deformation which has been previously cited as earthquake destruction/damage has 'been impelled not by settlement of seismic vibration but by the wanton and random smashing of underpinning courses for theft of iron and lead used in clamping.' (Cooper 1996, p 130)

**Current View of the Mechanism of Destruction:**
*Mechanical non seismic factors.* Possible earthquake damage in the past (i.e. Mediaeval period) demonstrated by offset column drums, but currently the monument has not been effected by large magnitude seismic events with epicentres in close proximity. Additionally, the slow, gradual disintegration is due to a host of factors, not least the mechanical and geological properties of the rocks and the effects of the prevailing extreme weather at the site (i.e. temp fluctuation of $13°$ C in the summer)

*Conclusion*
This small comparative study which can be considered representative of the assignation of destruction mechanisms in archaeological sites, demonstrates how the use of a 'identification criteria' or observation check list can lead to a misleading and at worst incorrect interpretation of the observed damage. Further, preconceived ideas of the expected effects of large magnitude earthquakes on, what are considered, fragile archaeological structures are shown to influence interpretation of the observations. This is clearly demonstrated with the Temple of Epicourion Apollo, Nemea, where the official reports following the 1986 Kalamata earthquake, whereas workers actually on the monument at the time of the event report were apparently 'unaware' of the earthquake and the monument experienced no observational damage. Clearly at this temple other adverse factors are contribution to the slow disintegration of the structure, which would be completely overlooked if a non specialist were to assign a

mechanism purely through the application on of an 'observational checklist'. Additionally, the observations of characteristic damage can also be the result of non-seismic mechanisms (Karcz & Kafri 1978), and this simple study demonstrates that it is important to construct an archaeoseismological appraisal within an interdisciplinary and holistic approach.

The next subheading in this section returns us to the third area of discussion in the theoretical concerns, namely communication.

### 3.4.4 Communication

Greater communication between practitioners of the contributing disciplines has been noted in all the archaeoseismological methodology papers as a way to increase the reliability of the resulting data (Rapp, 1986; Nikonov, 1988; Guidoboni, 1996). But what exactly is meant by greater communication? Even in a situation where a researcher is familiar with, has a close-relationship with, or has been instructed in all the contributing disciplines (archaeology and seismology in the case of this research), establishing cross discipline communication is not easy. There are two inter-related reasons for this. First, there are fundamental differences in philosophies and practice as highlighted in section 3.4 above, which discusses subjectivity and objectivity. Secondly, there are problems with the very basis of communication, language, which can be either explicit about the conflicting definitions of the same words or so subtle as to render communication more apparent than real (van Andel, 1994 p 26 – 27). Clearly, without real communication there can be no true collaboration.

### 3.4.5 Multidisciplinary/Interdisciplinary

As noted in chapter two, increased collaboration was advocated as a means to reduce the subjectivity in the identification of earthquake damage in archaeological remains. Van Andel (1994) discusses two modes of collaboration whose descriptors are commonly regarded as equivalent and used interchangeably. The first, and the one most prevalent in archaeoseismological studies, is multidisciplinary. This can best be described as research that involves contributions or data from more than one discipline, and is the type of collaboration that develops as the project moves forward and 'gaps' in the dataset are identified or archaeological discoveries are made. For example, an earthquake geologist studying a particular fault discovers in the ancient or historical literature a description of what appears to be a seismic event that affected the region in which the fault under study is located. Without any further investigation into the integrity of the data, the derived parameters (usually date and intensity) are incorporated into the seismological research and the results are published. The information is then perceived by subsequent workers as factual because of its inclusion in a 'scientific' peer reviewed journal and it is reused with no recourse to the context of the original data. This may seem an oversimplification, but examples can be found throughout the literature, for example the inclusion of apparent 'fake' or 'rogue' earthquakes in most respected earthquake catalogues; see (Buck and Stewart (2000) for a discussion of the seismogenic source of the 426 BC earthquake in Atalanti, Greece, (Bellettati, *et al.,* 1993) for the identification of 'fake' earthquakes in Italy, and (Ambraseys & White, 1996) for the Mediterranean region.

*Interdisciplinary* is the second mode of collaboration discussed by van Andel, (1994), and is described as 'collaboration which assumes intensive exchange of information, ideas and procedures from the planning stage through to final publication'. As the description suggests, achieving true interdisciplinary research requires maximum real communication between specialists carrying out separate investigations. As van Andel (1994) notes, this can only be realised by asking the following fundamental questions *prior to* and *throughout* the research project.

- What are, conceptually and operationally, the questions I should like some scientists [archaeologists/historians] to help answer?

- Which discipline or subdiscipline is best equipped to deal with these matters?

- How should my problem be phrased so that collaboration may be maximised?

- What do I need to learn so that I may understand the scientists [archaeologists/historians] answers and make maximum use of them?

- Can I afford this financially [with respect to the overall objective of the project]?

(van Andel, 1994)

Although these questions were originally formulated for archaeologists considering scientific collaboration, it would be equally beneficial if they were asked by scientists seeking archaeological or historical collaboration. For example, seismologists and earthquake geologists requiring information or advice from archaeologists and/or historians; to this end I have added the words in square brackets to the original questions.

As discussed above, the methodological papers in archaeoseismology and early research in historical seismology have demanded greater communication between the contributing (sub)disciplines as a means of reducing the subjectivity inherent in monodisciplinary interpretations of diverse datasets. It is clear that in order to obtain such dialogue it is necessary to design and execute true interdisciplinary research, with all participants understanding, or at the very least being aware of, the materials and data with which they are working. For example, the dating laboratory needs to be aware of the mode of collection of samples sent for analysis and the sampler needs to be aware of the procedures that will be performed upon the collected sample to ensure that the sampling technique fulfils the testing requirements. Additionally, the project co-ordinators (and it is suggested here that one from each principal contributing discipline is included in a project management team) need to be aware of the theoretical and practical limitations of the data provided by each contributing discipline. As noted by van Andel (1994), this places high demands on the collaborative research projects, but one which will reap the benefits of a more than superficial integration of theory, leading to the maximum use of, and benefit from the collected data.

There is still a lack of truly integrated cross-disciplinary research specifically in archaeology, even within supposedly interdisciplinary/ multidisciplinary projects, and here I use the terms interchangeably as there is little indication in the current literature (with the occasional exception such as van Andel (1994)) of the appreciation of the subtle difference. The cause of this may be 'the fear that interdisciplinary approaches can degenerate into an exercise in dilettantism' (Trigger, 1989. p 356). But, more realistically, it is probably the heavy demands on the project, in terms of time and money, which are the root cause. As a consequence, often the results of scientific analysis or historical investigations which are carried out in a multidisciplinary style are not fully realised and are simply tacked on to the main research report either as an annex or subsection, for example (Goodchild, 1968) and (Rapp & Gifford 1982).

Whereas true interdisciplinary research can provide a wealth of information to assist in, and provide context to, interpretations of collected empirical data, it cannot, as a proposed universal methodology, eliminate subjectivity. The reason behind this is that archaeoseismology lies at the interface between the human, social and natural sciences, and therefore developing a research methodology requires the understanding and integration of two distinct philosophies and practices. This is obviously difficult for any single individual who has been trained, or is experienced, in a particular discipline or methodological practice, and where cross-disciplinary training has been acquired it is difficult to maintain a balance between the individual disciplines.

If subjectivity is inherent within interpretation and cannot be completely removed, and if interdisciplinary, as opposed to multidisciplinary, research provides for maximum benefit from the derived data, it follows that a workable and useful strategy for any interdisciplinary research should include both points. Rather than concentrate upon refining a list of observational criteria, it is demonstrated in this research that a way forward is to construct a reasoned, and therefore rational, appraisal of the available data in light of an initial question and a method for the quantification of the uncertainties.

In summary, it can be said that the goal of any research is essentially the construction of a rational argument. In accordance with the suggestions by van Andel (1994) it is suggested that where the research involves multiproxy data, i.e. interdisciplinary collaboration, the construction of universal methodologies or criteria inhibits the bi-directional feedback necessary to achieve this goal. It is the aim of this study to demonstrate the value of a truly interdisciplinary approach by using a tailored strategy using the template proposed in Figure 3.1 as a replacement for the traditionally held view of 'descriptions' or 'criteria' as a universal methodology.

## 3.5 Identification of Data Sources

Interdisciplinary research will principally draw on two different data sources, both of which are equally legitimate if the project leader (co-ordinator) considers both the context and limitations of any information obtained. The first are indirect sources and essentially provides information derived from established literature, pre-existing publications or results of primary fieldwork carried out by collaborative partners (either for this or alternative research projects). It follows, then, that information obtained through direct fieldwork or analysis of the empirical data (i.e. observations or objects) is the second source of information, which can also be described as primary data.

A fundamental precondition for interdisciplinary research is that that the project co-ordinator (or research leader) will find all but the data pertaining to their own specialised field from indirect data sources. Another is that the primary data collection cannot be effectively planned or carried out until the 'gaps' in the empirical data base have been identified, through analysis of the derived data sets in light of the data layers identified as essential to the investigation.

Essentially, the pulling together of multiple and diverse datasets allows the construction of a corpus of information which forms the empirical database from

which further analysis takes place. Accessibility and manipulation of derived data is the key to the format of the contributing data, and it is suggested that where applicable a chronological order should be used as it facilitates useful comparison and or correlation. For example, understanding intermittent occupation of coastal sites may be possible from a comparison of environmental data (i.e. information about relative sea level changes) with both physiographic data (i.e. topography and localised seismic activity) and the archaeological historiographical interpretations of specific sites.

A well-constructed empirical database is the result of good interdisciplinary research, and is essentially the accumulation of information relevant to the research strategy. As each data layer is the result of discrete investigation, it should also be able to stand alone as an individual empirical dataset for monodisciplinary research and be amenable to cross-discipline consultation for other multidisciplinary projects.

## 3.6 Feasibility

Project feasibility is specifically the assessment of the probability of the research strategy reaching the set goals or answering the initial question. In well-planned interdisciplinary projects the probability of successful completion in terms of the theoretical issues will be high. However, the practicalities also need to be addressed, and an integral part of total project completion is data collection and/or field testing of theories and/or hypotheses. There are two elements to the feasibility of data collection and field testing, firstly whether the proposed dataset will make an essential contribution to the project, and secondly the selection of a suitable, and unbiased, field area (and this term is taken to include any data collection point, for example the natural environment, library, records office, etc.) that will provide the desired data sets. In truly interdisciplinary research the project co-ordinator (or management team) should continually assess the feasibility of collaborations and resulting data sets (see multidisciplinary/interdisciplinary section above).

In order to fulfil the aim of this research, to determine what archaeological remains can tell us about palaeoseismology, it was necessary to carry out an archaeoseismological appraisal using a field area selected on the basis of the following criteria.

- The presence of a geographical and temporal range of previously excavated, and published, archaeological sites in close proximity to a known seismically active structure.

The Aegean region is one of the most seismically active regions in the world and one that fulfils the archaeological requirement of having a long and relatively well-documented history of human occupation. Due to the strict regulations governing archaeological data and fieldwork the final selection was guided by the following criteria:

- The maximum number of archaeological sites, with previously published literature, located near to a single seismogenic structure.

The Atalanti region in central mainland Greece fulfilled both criteria and additionally contained archaeological sites which had previously been used in archaeoseismological research. It therefore holds the potential to fulfil the aims of the interdisciplinary research methodology proffered here.

## 3.7 Methodology Applied to this Research

The previous subheadings in this chapter have concentrated on discussing the theoretical issues associated with establishing an effective interdisciplinary research project with a clearly defined project objective based upon the end user of any derived information or results. The remaining subheadings outline how this new project-specific methodological approach has been applied to this research. This discussion starts from the theory-methodology transition stage (where methodology is defined as a collection of procedures or processes aimed at data acquisition) through to the interpretation in the final stages of the research project.

### 3.7.1 Theory/Methodology Transition Stage

Defining the research question (i.e. the refined question) is simply a process of development of the research strategy through making definite and informed choices. In principle this should be carried out throughout the theoretical section planning to ensure that strategy, and ultimately the practical methodology, will provide the necessary data for the interpretative stage. Although this is rarely considered as a discrete element of research it tends to be omnipresent and can be seen most clearly when, after the literature review or background enquiries in a general area of interest the field of is narrowed down to a discrete topic (or issue) as the primary focus of the project for further investigation. This often takes the form of a question to be answered or specifically in the geological discipline the application of an existing method of investigation to a specific and well-defined field area.

The background research and literature review for this project raised a number of diverse and quite disparate key issues, each of which individually could potentially form the basis of the refined question. However, as discussed above, the key to successful interdisciplinary research is to remain focused on the needs of the ultimate end user especially throughout the key conception stages of the strategy where early tangents can jeopardise the future

interpretations through the collection of miscellaneous and/or inappropriate data. In this project it is the potential of archaeologically derived palaeoseismic data, with considerations throughout the strategy concentrating on the earthquake geologists as the major end users being the is the focus with the benefits to the archaeological community taking a secondary position. Having defined both the end user and the aims of the project, this was translated into the following refined question from which to construct an applicable practical methodology.

*Are archaeological sites and/or remains a useful proxy data source for palaeoseismological research?*

This question not only allows for consideration of both theoretical and methodological issues (i.e. subjectivity, feasibility/practicalities of interdisciplinary research, and construction of an empirical database), but through the selection of a field area that contains previously published supposed archaeoseismic damage it allows for a comparative analysis which can be considered the crux of the research project. The key issues considered throughout the formulation of the research strategy used in this work are shown in diagrammatic form in Figure 3.4. Note that this figure deals only with the first two stages of research, i.e. theory and methodology, which are discussed in this and the previous chapters. The construction of the empirical data based from culled data is discussed, where necessary, in chapter four.

### 3.7.2 Data Collection

Data collection falls in the methodological part of the research outline (Figure 3.1) and consists of two parts: (1) defining the data needed for the research in terms of actual data sets required, and (2) defining the methods of acquisition. As noted in section 3.5, there are two principal sources of information, direct and indirect, for both the archaeological and geological elements.

*Direct Data*
Direct sources describe primary observations (archaeological and/or geological) and the means of collection are principally site and field work using standardised practices, i.e. geological and geomorphological mapping, archaeological site observations etc. In all cases a field notebook was kept and an extensive field photograph collection has been amassed. The photographic collection proved a particularly useful reference as extensive road maintenance and upgrade has completely obliterated some prime outcrops.

Geological mapping was carried out by systematically walking the field area looking for bedrock outcrops. Geological measurements were taken using a standard Sunto compass/clinometer and macro observations, including dip and strike measurements and lithology, were pencilled on field slips of the 1:10,000 Hellenic

Army Geographic Service (HAGS) topographic map. These were then later checked against the existing Institute of Geology and Mineral Exploration (IGME) geological Map (Atalanti Sheet).

Geomorphological observations were restricted to the vicinity of the fault and were made in conjunction with the 1:30,000 geotechnical maps from Angelides (1992), although a detailed slope study was carried out on one lower fan deposit behind the Lowenbrau beer plant at Kyparissi. This was carried out using an Abney level and was conducted as part of the Kyparissi site case study (see chapter 6). A detailed assessment of the unconsolidated Quaternary deposits was not possible due to the specialised nature of the analysis required, and in this instance observations by Angelides (1992) were accepted. Coastal observations were made; approximate positions of features, both geological and archaeological, were again noted on field slips of the 1:10,000 HAGS topographic maps.

Archaeological fieldwork was restricted to by permit to observations only, except for Allai where Prof. John Coleman granted permission to survey remains within the delineated boundaries (and excluding the coastal area). However, the site had been extensively surveyed the previous excavation field seasons (1986 – 1996) and partially surveyed in association with the 1997 excavation season, therefore it was decided that it was not necessary to duplicate the work with another full survey.

*Indirect Data*
Essentially these are secondary sources, i.e. where the observations have been made by another party and are published and/or made available to the project. In true interdisciplinary research all but the project co-ordinator's own field of expertise will have an indirect origin and will be culled from a range of source material. Methods of collection included trawling literature such as newspapers, reports and classical texts, and scientific reports. Using indirect sources requires a starting point for investigation, for example the known date of an earthquake or the name of a site that may contain earthquake damage. Trawling the available data sources without a 'lead' renders this particular method of data collection time-consuming. Additionally, once a reference is located, the useful information may amount to only a single word, date or paragraph of description.

Due to the paltry nature of many of the historical and cartographic references used in this research, the data were simply collectively tabulated (see Appendix 3) and brought into the discussion where applicable.

Other indirect sources include reports and contributions of other specialists working for or affiliated with the research project. As these contributions will have been designed for the benefit of the project the reports should contain the results in a usable format. For example, if an archaeological survey team has been commissioned to

provide a spatial analysis of the archaeological sites in the region, this should ideally be presented in map format, with a separate discussion.

### 3.7.3 Interpretative Stage

The interpretative stage essentially comprises three levels, the results of which are derived from the level below and contribute to the level above (see Figure 3.1).

*Low Level Theories*
The first stage in interpretation is to construct low level theories (Trigger, 1989). These can be described as empirical research with generalisations, and are characterised by propositions driven primarily by observations drawn from the collected data (Trigger, 1989). Trigger notes that the inferences in this stage can be described as specific attributes (or in the case of archaeology, artefact types) which occur in a particular location both individually and in association with each other.

In essence this represents a purely inductive approach where 'presuppositionless observation is intended to illicit the true nature or cause of a particular phenomena' (O'Hear, 1989). However, the process of constructing the initial question through background research and theoretical considerations informs the research and effectively negates the idea of presuppositionless observation. Therefore, it is better to consider the extraction of generalisations at this low level as the construction of a testable 'working hypothesis'. This shift away from a purely inductive methodology not only allows for the imposition of bold, but apparently weakly supported (in terms of empirical data) statements, such as "this site was destroyed by earthquakes in the era before Christ", but also prevents the laborious stock-piling of irrelevant observations Thus, in the case of this thesis, the repeated observations of damaged remains/relics in archaeological sites in close proximity to known seismogenic sources constituted the working hypothesis that these sites can be used as indicators, or 'barometers', of local seismic activity. In the next stage this working hypothesis was tested using a control environment and by reducing the variables such that a value judgement could be established.

*Testing the Working Hypothesis*
Previous methodologies have concentrated on the assessment of a site for seismic damage through the use of identification criteria or a checklist approach, which has often resulted in a *deus ex machina*[14] interpretation assigning seismic mechanism to the observed damage. This approach gives little or no consideration of the geological aspects, and places emphasis on the intra-site[15] observations meeting the 'earthquake damage criteria'. Essentially the field observations coupled with the

---

[14] See footnote 3
[15] Site specific

29

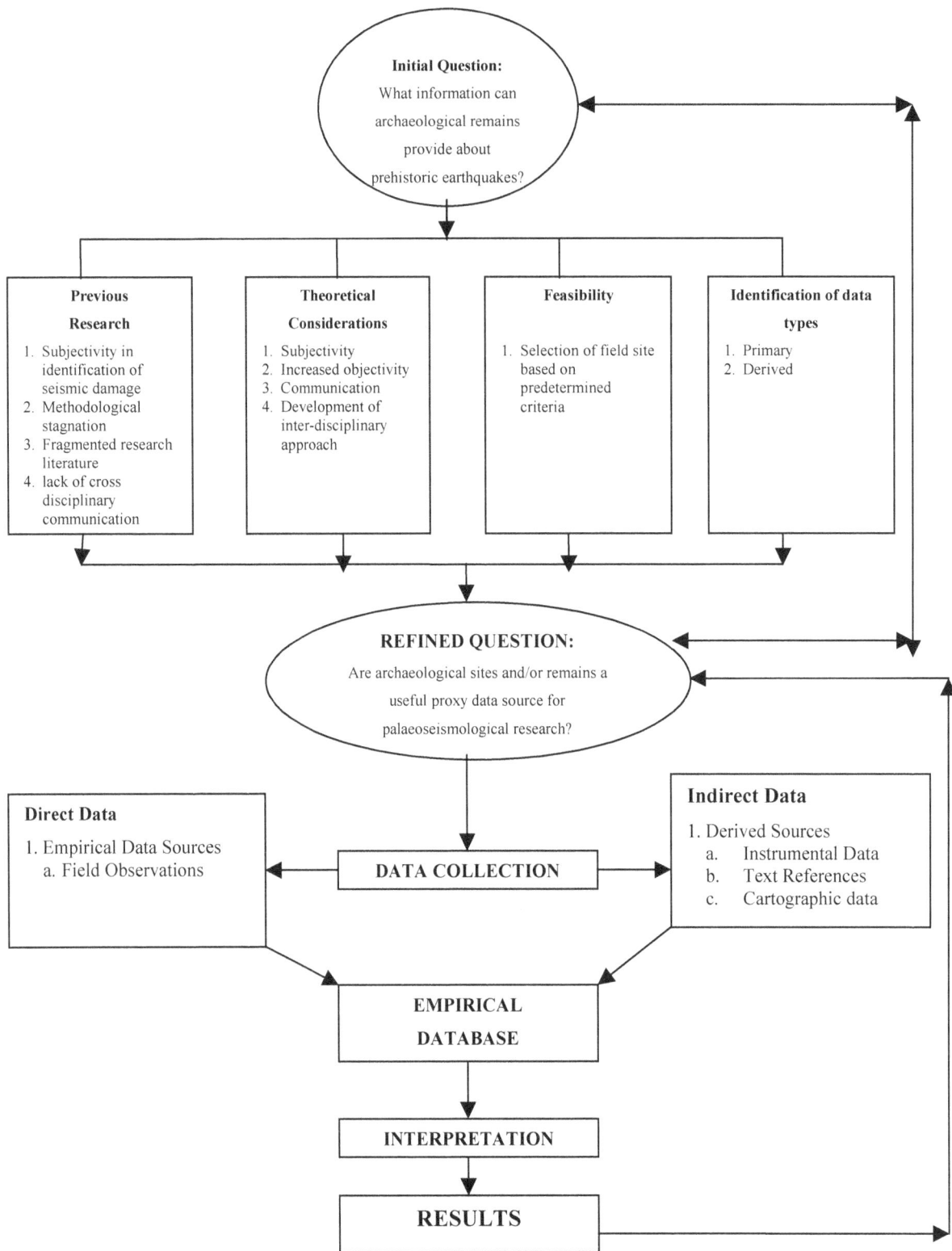

**FIGURE 3.3**: FLOW DIAGRAM SHOWING THE CONSTRUCTION OF THE RESEARCH STRATEGY USING THE METHODOLOGY OUTLINED IN THIS CHAPTER. EMPHASIS HAS BEEN PLACED ON THE THEORY AND METHODOLOGY SECTIONS OF THE PLANNING.

interpreter's preconceptions, inform the interpretation. This is an entirely unsuitable, and highly biased, base from which to make a value judgement, such as is required for this research, as it is precisely the field observations of the cultural remains which are being assessed as quality indicators of seismic events. Furthermore, alternative explanations (natural and anthropogenic) are rarely, if ever, considered. It is, therefore, necessary to adopt an alternative approach.

In this research it is suggested that, rather than using the cultural remains to derive the geological event, and thus making a value judgement on the highly subjective interpretative data, the reverse should be carried out. Thus, by using modelled geological parameters it is possible to make a comparative analysis of what is expected geologically, and what can be observed in the cultural remains. By using this method it should be possible to test the working hypothesis that archaeological remains are useful indicators of seismic activity by comparative analysis of supposed seismic damage. A Geographical Information Systems (GIS) program was identified as a tool to facilitate the comparative analysis due to the spatial nature of the geological models and the supposed seismic damage in archaeological sites.

*Results/Interpretation*
The final stage of the interpretation, and of the research strategy as a whole, is the presentation of the results or interpretations. In this thesis these take the form of a written discussion, but they could in other circumstances also include tables and graphs. The advancement of research demands that results will usually give rise to new questions and that as shown in the flow diagram at Figure 3.1, these new questions will form the basis (initial questions) of a new research strategy.

# Chapter 4: Atalanti Database

The information contained in this chapter forms the empirical database for the field area (Atalanti, central mainland Greece) set out in the previous chapter 3 (Theory and Methodology). This chapter can be roughly divided into two distinct parts based on the requirements of the research objective. The first comprises data pertaining to the seismological element of the research and includes information on the physical environment with emphasis on the tectonic regime and geological setting. The second part essentially deals with the anthropogenic factors, principally drawing from the published archaeological and historical records. Due to the restrictions of the archaeological research permit only three sites within the field area (Kyparissi, Alai, and Kynos) have been discussed here.

The order of presentation (and where necessary, discussion) within the chapter is independent of the sequence of collection, and the division of the data into the two parts is maintained solely to provide the reader with a clear presentation of the information. Further, despite the majority of the datasets being identified prior to the collection stage, desiderata were gathered as a result of the zetetic nature of the research strategy. Therefore, it is in reality the end result of the identification and accumulation of data required to fulfil the research aims, and can be thought of as an outgrowth or product of the research project. As noted in the discussion of chapter three, this attaches an external value to the database beyond this particular research project and the specific research objectives and questions posed here.

## 4.1 Background to Atalanti/Opuntian Lokris Region

The proposed study area (Lat. 38° 35' - 38° 45, Long. 22° 30' - 23° 30'), shown as the boxed area in the location map at Figures 4.1 and 4.2, is part of the Division of Lokris (Επαρχια Λοκριδος) in the Province of Phthiotis (ΝΟΜΟΣ ΦΘΙΩΤΙΔΟΣ) and focuses on the Atalanti Plain and the surrounding coastline located on southern margin of the Gulf of Evvia.

The field area is bordered to the north by the shores of the Bay of Atalanti and is divided from the rest of mainland Greece to the south by the Khlomon mountain range (max. elevation 1080m). The Atalanti plain (approx. 10km wide), which can be considered the centre of the field area, is virtually isolated from the rest of mainland Greece except for two coastal passes. The narrower of these lies to the south-east of the plain, where the waters of the Bay of Almyros are separated from the base of the mountains only by the raised causeway of the National Road (E75). The north-eastern pass is not as pronounced due to the more gentle climb up to the Xerovouni plateau.

The fertile plain of Atalanti extends approximately 8km from the foothills of Mount Khlomon in the south to the Gulf coast, where the views of the bay are dominated by the presence of two large bare rocky islands, Atalanti Island (Αταλαντη Νησι) to the west, and Donkey Island (Γαιδαρος Νησι) to the east (see the 1:50 000 topography map extract in Figure 4.2 for the location of the islands in respect to the plain). Olive groves, and to a lesser extent tobacco, are grown on the lowest slopes of the Mount Khlomos range front, whilst market gardening and horticulture predominates closer to the coast, with salt marshes occupying much of the immediate coastline in the south east (Almyra) area. To the north west, salt marsh gives way to shingle beach, due to the slight increase in elevation beyond Atalanti Island.

The general topography described above can be clearly seen in the satellite image of the field area at Figure 4.2. The Atalanti plain is visible to the centre left of the image and Atalanti town is indicated by the letter A. A sharp topographic contrast between the plain and the Khlomon range trending diagonally across the image (and highlighted by the red line marked AF), marks the generally accepted trace of the Atalanti Lokris fault which approximates to the Quaternary/bedrock contact (IGME, 1965). The topographic/physiographic setting of Atalanti (i.e. juxtaposed topographic highs and lowland plains) is characteristic of central Greece, and is the physiographic expression of the localised crustal deformation on individual normal faults which accommodate the prevailing regional scale extensional tectonic regime in the Aegean.

## 4.2 Tectonic Setting

The complex tectonic and geomorphological framework of the Aegean has evolved as an integral part of the Mediterranean region as a whole. It is directly linked to the palaeogeography of the Mesozoic Tethys ocean and its' subsequent closure due to the collision of the African, Eurasian and Arabian plates in the Late Mesozoic - Early Tertiary (Robertson & Grasso, 1995).

The geology of the Aegean region (which lies in the eastern part of the Mediterranean) is traditionally described in terms of groups of rocks (Isopic Zones) that share a common history, both in terms of the original sedimentary depositional environment and structural characteristics, and the ancient metamorphic and plutonic rocks (Massifs). A pre-existing structural fabric can be observed in the geology of Greece as the result of complex palaeotectonic faulting (thrust) and folding associated with Alpine and Aegean geological history and is characterised by the presence of stacked groups (or nappes) of rocks (Dermitazakis, 1984). This has subsequently been overprinted with the complex neotectonic regime which is discussed in section 4.2.2.

**FIGURE 4.1:** OUTLINE LOCATION MAP OF THE AEGEAN. THE FIELD AREA IS MARKED BY THE BLACK DOT AND IS EXPANDED BELOW IN FIGURE 4.3

**FIGURE 4.2**: SATELLITE IMAGE FROM THE LANDSAT 4 TM (28 JAN. 1988) ENHANCED CHANNEL 5 (SHORT-WAVE INFRA-RED) RAW DATA, PRE PROCESSED BY A. GANAS (1998). THE ATALANTI PLAIN IS VISIBLE TO THE CENTRE LEFT OF THE IMAGE, ATALANTI TOWN IS INDICATED BY A, DONKEY ISLAND BY D AND ATALANTI ISLAND BY AT.

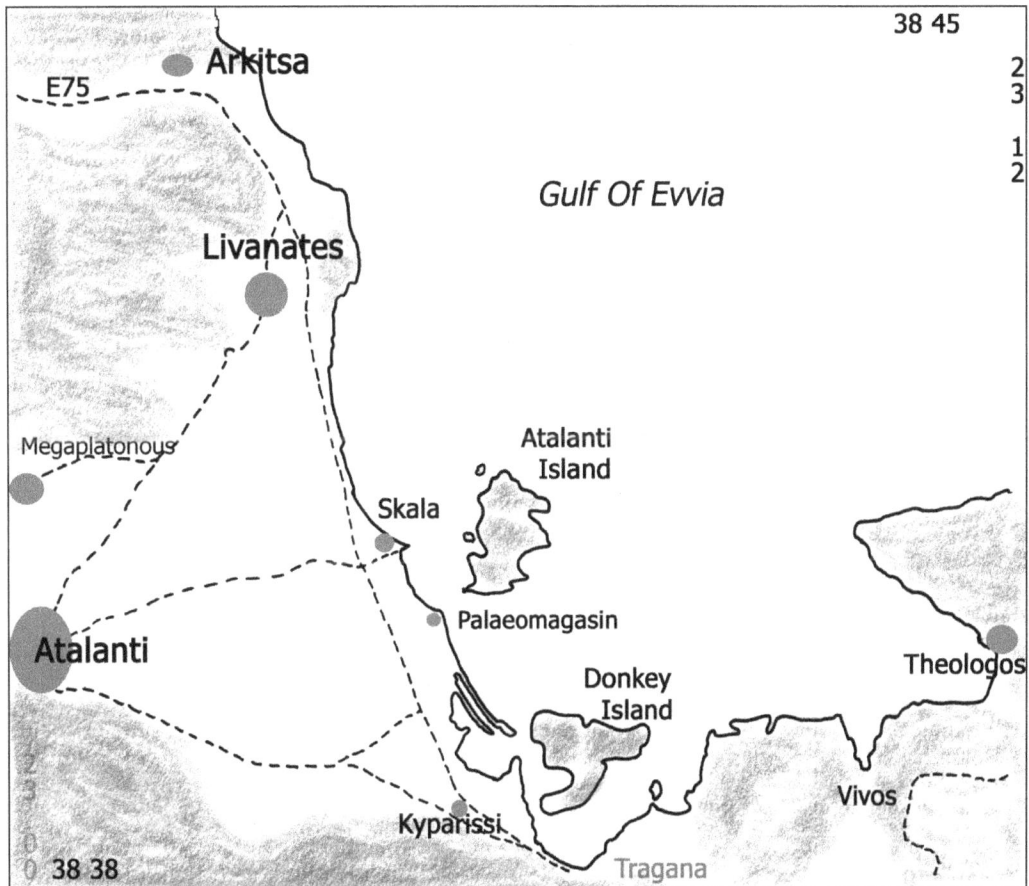

**FIGURE 4.3:** SKETCH MAP OF THE 1:50,000 HAGS TOPOGRAPHY MAP OF THE ATALANTI FIELD AREA LAT. 38° 38' – 38° 45', LONG. 23° 00' – 23° 12'

### 4.2.1 Isopic Zones and Massifs

Central mainland Greece is characterised by number of NW - SE trending isopic zones (Dermitazakis, 1984), with the field study area straddling the middle of the Pelagonian and Sub-Pelagonian zones (Higgins & Higgins, 1996). Shallow water limestones characterise the Pelagonian (e.g. the deposits north of Theologos on the Malesina peninsular), while ophiolites (and associated rocks such as deep water limestones and cherts) dominate Sub-Pelagonian. The transition between the two zones can be observed in the juxtaposition of ophiolitic deposits with limestones in the coastal outcrops north of Tragana and the footwall of the Atalanti fault south of Kyparissi.

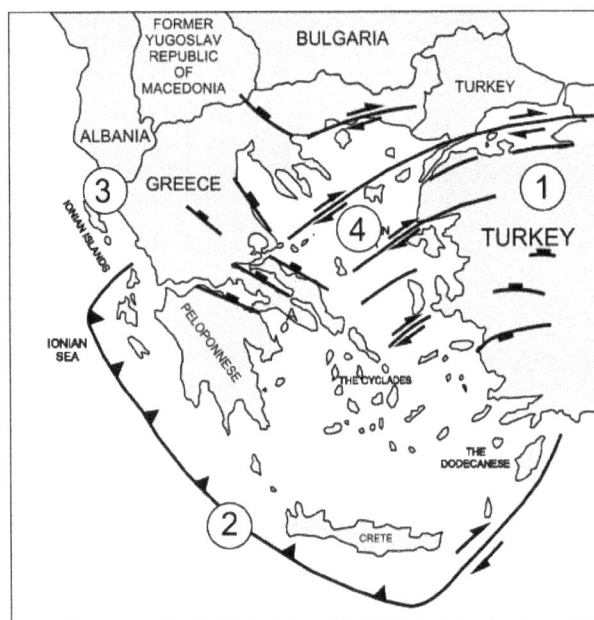

FIGURE 4.4: THE TECTONIC PROVINCES OF GREECE AND SURROUNDING AREAS. PROVINCE: (1) WESTWARD EXTRUSION OF ANATOLIA; (2) SUBDUCTION AT THE HELLENIC TRENCH; (3) CONTINENTAL SHORTENING IN WESTERN ALBANIA AND THE NORTH-WEST COAST OF GREECE; AND (4) NORTH-SOUTH EXTENSION OF CENTRAL AEGEAN. (NOT TO SCALE).

### 4.2.2 Neotectonic Framework

Although, at present, there is no universally accepted explanation for the neotectonic[16] deformation geometry and rates of extension in the Aegean region, a variety of models have been proposed over the past three decades.[17] The main focus of these models has been to account for the diverse geological, topographic and seismological observations which occur within the relatively small area occupied by the Aegean and which can be described in

terms of tectonic provinces (see Figure 4.4). Below, the main tectonic provinces are briefly noted.

*Province 1*: the westward extrusion of Anatolia, as a result of the African-Eurasian plate convergence (Jackson, 1994), is accommodated by right lateral shear along the North Anatolian Fault (NAF) which terminates in the far west in the North Aegean Trough (NAT).

*Province 2*: subduction of the African plate beneath the Aegean plate at the Hellenic Trench is manifest in the concentric arcuate features of the Hellenic Island Arc (indicated by the islands Kephalonia-Crete-Rhodes), the South Aegean Trough, and the Cycladic volcanic island arc marked by the islands of Milos (noted as one of the primary sources of obsidian in antiquity), Santorini (Thera) and Nisyros (still showing volcanic activity).

*Province 3*: active continental shortening along the western coast of Albania and the far north-west of Greece is characterised by earthquakes with reverse and thrust fault mechanisms (Jackson, 1994).

*Province 4*: active north-south extension of the central Aegean is accommodated along large normal faults which form the characteristic NW-SE grabens that dominate central mainland Greece and the NE-SW trending normal faults of western Turkey.

### 4.2.3 Regional Tectonic Models

The basis of all broad scale tectonic models of the Aegean starts from the initial work of McKenzie (1970; 1972), who derived the plate boundaries and motions in the Aegean region from seismicity and focal mechanism solutions (see Figure 4.5). It should be noted, however, that there is no evidence for the connection of the North Anatolian Fault with the Hellenic Trench by a large strike slip zone, and this specific feature of the original model is now considered to be incorrect.

More recent models fall into two main approaches: the continuum mechanics of Jackson and colleagues (Taymaz et al., 1991; Jackson, 1994) (Figure 4.6), and the strain localisation of Armijo and colleagues (Armijo et al., 1996) (Figure 4.7). Both approaches start with and agree on the following points:

- the westward extrusion of the Anatolian block due to the continued convergence of the plates in the Eastern Mediterranean;
- the subsequent accommodation of this movement along the North Anatolian Fault (NAF) and East Anatolian Fault (EAF);
- the approximately northward subduction of the African plate at the Hellenic subduction zone.

They differ, however, in regard to the timing and order of development of the seismogenic structures that dominate central mainland Greece, as the following outlines of the

---

[16] Neotectonic is defined by Hancock and Stewart as 'a branch of tectonics concerned with understanding earth movements that both occurred in the past and are continuing at the present day' (1990, p 370 & 371).
[17] The reader is directed to Armijo et al., (1996) and Jackson, (1991;1994) for extensive review and critique of previous work.

two models will highlight. Detailed discussions of the individual models can be found in the primary sources of Jackson (1991; 1994) and Armijo *et al.* (1996).

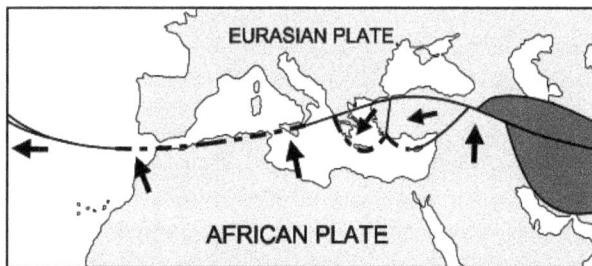

**FIGURE 4.5**: MCKENZIE'S (1972) INTERPRETATION OF THE SEISMICITY AND FOCAL MECHANISM SOLUTIONS FOR THE AEGEAN REGION WHICH HAS FORMED THE BASIS OF ALL THE SUBSEQUENT MODELS. NOTE THAT SINCE THE PUBLICATION THE INTERPRETATION OF A STRIKE SLIP FAULT ACROSS THE NORTH OF GREECE HAS BEEN DISCOUNTED DUE TO THE LACK OF ANY GEOLOGICAL EVIDENCE (REDRAWN FROM MCKENZIE, 1972).

*Broken Slat Model (Figure 4.6)*
This model states that the continued collision of the African, Eurasian and Arabian plates has led to the tightening of the African - Eurasian suture and crustal pinching, resulting in the westward extrusion of the Anatolian block (Taymaz *et al.*, 1991). As a result of the continental contact between the Aegean and Anatolian plates, the Aegean plate migrates southwest, overriding the African plate along its southern margin, the Hellenic Arc (McKenzie, 1970). Subduction of the African plate results in the arc volcanics of the Southern Aegean Sea (Cycladic Islands), and in back arc stretching (extension) in central Greece and western Turkey driven by 'roll back'[18] of the subducting slab (Taymaz *et al.*, 1991).

The principle movement direction along the EAF is ambiguous, but the available data suggest that motion is predominantly left-lateral strike slip along a diffuse zone with an element of shortening (Jackson, 1994). The sense of motion on the NAF is accepted as right-lateral strike slip along a single, well defined fault that splays into a number of sub-parallel strands across the North Anatolian Sea, trending NE-SW beyond the 30° longitude (Jackson, 1994). The strike slip faulting associated with the NAF does not cross central Greece, but terminates in the far western end of the North Aegean Trough abutting the large NNW-SSE trending normal fault-bounded grabens (see Figure, 4.5) which have SSE - SSW slip vectors (Jackson, 1994).

Jackson and McKenzie (1994) note that the switch in the slip vectors and strike of the faults within the mainland, coupled with the abrupt change in the strike, suggests that the right lateral shear is accommodated by clockwise

rotation of the fault bounded blocks in central mainland Greece. Palaeomagnetic studies by Kissel and Laj (1988) appear to substantiate this inference (Jackson, 1994), and Taymaz *et al.* (1991) suggested a simple broken slat model to explain the apparent motions on the faults.

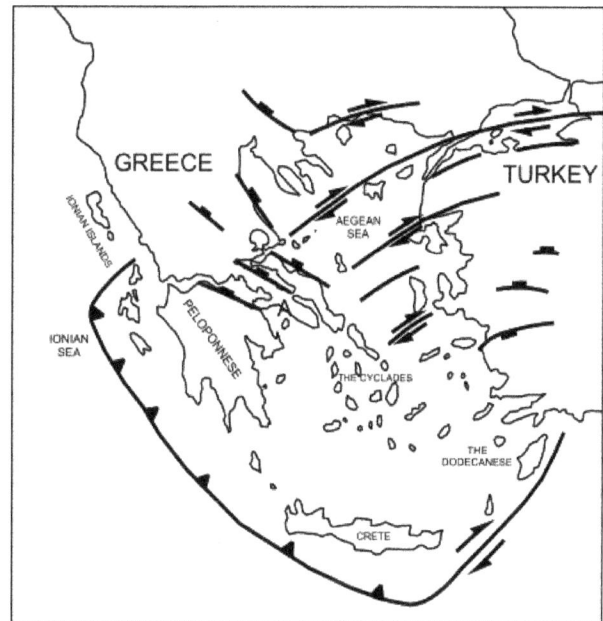

**FIGURE 4.6**: OVERVIEW OF THE 'BROKEN SLAT' TECTONIC MODEL PROPOSED BY TAYMAZ *ET AL.*, (1991).

The model describes the observed motion on both the strike slip and the extensional faults (derived from focal mechanism solutions) throughout the Aegean Basin by the rotation of two sets of slats (representative of the fault bounded blocks of mainland Greece) in opposite senses. The rotation of the western set (mainland Greece and, Peloponnese) is almost double that of the eastern set (western Turkey and Eastern Aegean Islands), with increased rotation towards the south to account for the curvature of the Hellenic arc (Jackson, 1994). Satellite Laser Ranging (SLR) measurements appear to corroborate the calculated velocities, allowing for the fact that the non-geodetic values do not take into account the contributions of aseismic creep and earthquakes with magnitude less than 6 (Jackson, 1994).

*Strain Localisation Model (Figure 4.7)*
Following extensive research in the Gulf of Corinth Armijo *et al.* (1996) proposed a model to provide a "simple, unified interpretation of the kinematics of the Aegean from the Late Tertiary to the present". The model was developed to include all the significant features of earlier models by McKenzie (1972) and by Taymaz *et al.* (1991), and incorporates previously inconsistent data, such as evidence for an earlier phase of crustal extension (i.e. pre 7 - 10 Myr) and the observation that strike slip motion does not appear to be distributed throughout western Turkey on splay faults but is primarily accommodated on the North Anatolian fault (NAF)

---

[18] Roll Back refers to the process where by the subducting slab curls back on itself causing extension on the surface.

(Armijo et al., 1996). Additionally, the proposed model incorporates the fundamental concepts that (1) geological structures must evolve in a way that is consistent with present understanding of the earthquake cycle, and (2) faults evolve by a process of propagation, with small scale faulting restricted to the crust, while at larger scale rupture must involve the upper mantle (Armijo et al., 1996).

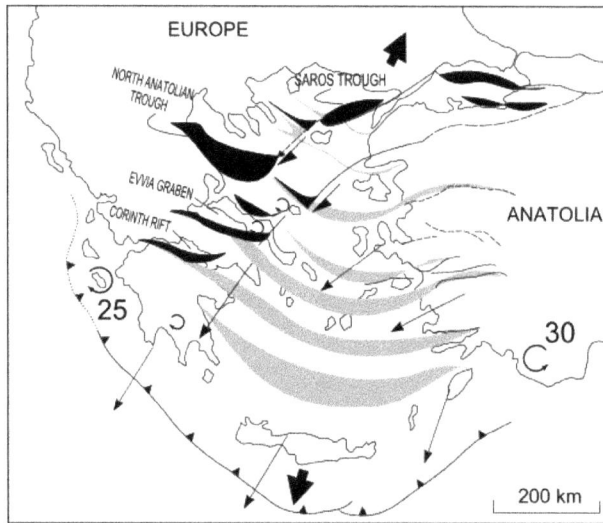

**FIGURE 4.7:** OVERVIEW OF THE 'STRAIN LOCALISATION' TECTONIC MODEL PROPOSED BY ARMIJO *ET AL.*, (1996).

As mentioned above, one of the key points of this model is that there are two phases of extension, with the early phase of crustal stretching in a roughly N-S direction resulting is localised lithospheric failure (Armijo et al., 1996) around 15 Ma (or earlier). This is followed by a second phase of stretching, beginning around 5 Ma (as a result of the development of the North Anatolian Fault plate boundary and the subsequent extrusion of Anatolia), with maximum extension in the centre of the Aegean Sea, decreasing towards the eastern and western margin, and resulting in arcuate rifting due to associated rotation both clockwise and anti-clockwise respectively. Notably, in this model, the North Anatolian Fault propagates into an active extensional environment, cutting and interconnecting pre-existing rifts which pull apart basins such as the Saros Trough, Sea of Marmara and North Aegean Trough. Additionally, the model suggests that the current seismicity should focus on the Corinth Rift which is corroborated in the instrumental seismicity records and through detailed mapping of the uplifted marine terraces located on the footwall blocks of the bounding rift faults.

In summary, and without prejudice for either model, Central Greece is dominated by large graben systems (up to 100km long) formed by large NW - WNW trending normal faults to accommodate the overall north - south extension in the Aegean. The Atalanti field area is located in central mainland Greece, and is dominated by a large active Quaternary normal fault that is part of the Gulf of Evvia rift system. Down-shifting the scale of observation, the preceding section outlines the geometry and deformation associated with active normal faults.

### 4.2.4 Normal Faults

The overall motion on normal faults consists of the relative down-throw of an overlying 'hangingwall' block, in the direction of the dip of the fault, with respect to the 'footwall' as the result of shallow (i.e. <10 - 20km), powerful (up to Magnitude 7) earthquakes. Fault plane dips range between 30 - 60°, though, as discussed by Jackson and McKenzie (1983) they may become curved (listric) at a depth of around 10 - 15km depth. Further, this geometry promotes rotation of the originally horizontal or flat beds, resulting in hangingwall depression (subsidence) and the development of antithetic structures (faults that dip in the opposite direction). In map view the fault traces of pure dip-slip faults are highly irregular, showing not only that normal faults occur in systems (or fault arrays) but that the individual faults within the systems are subdivided into sections or segments (Schwartz, 1988).

### 4.2.5 Normal Fault Segmentation

Field mapping following large-magnitude earthquakes shows that rupture does not occur on the full length of the fault, but only on certain segments (Schwartz, 1988). Jackson (1988), discussing the along-strike continuity of large faults, noted that even during the largest normal faulting earthquakes (up to 60km) surface rupture was never continuous and occurred on discrete sections or segments up to approximately 20km in length. Yeats *et al.* (1997. pp 282) and De Polo (1991) suggested that segments can be identified by the presence of discontinuities in the fault zone that fall into three major types: (1) structural (i.e. fault bifurcation or intersection with other faults), (2) geometric (i.e. changes in directions of strike, presence of *en echelon* stepovers), and (3) behavioural (for example, seismic behaviour, rates of rupture, sense of displacement, etc.). Of these, only the first two impede rupture propagation.

In areas of active crustal extension, normal fault segments can be identified by topography, i.e. "areas of the fault system with relatively large fault displacements (several kilometres) that lie between persistent segment boundaries which are marked by low fault displacements (tens of metres)" (Roberts & Koukouvelas, 1996. p 620). A distinction should be made here between segments as topographic expressions of active seismicity, within a fault system, and earthquake (or rupture) segments, which are the lateral extents of the ground rupture, or the portion of a segment that ruptures during a discrete seismic event and which can be variable along the topographic segment.

The Gulf of Evvia rift system is subdivided into a number of segments on the north and south edges, including the

Atalanti (Lokris) fault segment which dominates the bay and plain after which it is named. The palaeoseismicity of this fault segment is not well established, though it is known to have last ruptured in 1894 during one of the largest double event earthquakes (M = 6.2 and M = 6.9) experienced by mainland Greece (Ambraseys & Jackson 1990). There is no overall consensus about the dimensions of the Atalanti Fault; some workers claim lengths up to 70km (Stiros, 1988) or 40km (Ambraseys & Jackson, 1990), suggestions which appear to originate in the misinterpretation of the contemporary literature of the 1894 earthquake sequence (Papavasiliore, 1894; Skouphos, 1894; Mitsopoulos, 1895). Others propose smaller segment lengths, for example 34km proposed by Ganas (1997) and Ganas *et al.* (1998), based on remote sensed topography and stratigraphic observations. The data layers used in this empirical database make use of deformation and seismic hazard calculations derived from the smaller segment length of 34km proposed by Ganas (1997).

### 4.2.6 Earthquake Behaviour Models

Fault segmentation is the foundation upon which earthquake recurrence[19] and therefore fault behavioural models used in hazard assessments, are constructed. Central to these models is the concept of the 'earthquake cycle', which refers to the loading and relaxation processes associated with tectonic structures and the patterns of slip on the fault segments (Roberts *in press*, Schwartz, 1988):

*Characteristic Earthquake Model*
- Constant displacement per event at a point
- Variable slip rate along length
- Constant size large earthquake; infrequent moderate earthquakes

*Uniform Slip Model*
- Constant displacement per event at a point
- Constant slip along length
- Constant size large earthquakes; more frequent moderate earthquakes

*Modified Overlap Model*
- Constant displacement per event at a point
- Variable slip rate along segment length, where overlap includes full extent of the rupture, i.e. centres as well as segment ends
- Does not differentiate between the size and frequency of earthquakes.

Scholtz (1990) suggests that the characteristic earthquake and uniform slip models are essentially end members that will fit some case studies better than others, whilst alternative models that fall along this continuum will also fit specific case studies; for example, the recently proposed Modified Overlap model developed by Roberts (1996) using lineation data collected on the South Alkyonides fault within Gulf of Corinth rift system.

As noted above the characteristic earthquake model asserts that characteristic earthquakes (i.e. earthquakes of the same size and having a narrow range of magnitudes which are at or near the maximum possible) are generated by individual faults and/or fault segments. The model (proposed by Schwartz & Coppersmith 1988) assumes that the slip distribution associated with characteristic events is repeated in successive events, resulting not only in constant single event displacements at any given point, but also in along strike slip-rate variability.

Fault behaviour models have important implication not only for the seismological aspects of fault growth, understanding along-strike displacement variability (both in individual faults and large rift systems), and variable slip rates, but also for research that relies on an understanding of active faulting; for example studies that attempt to derive past neotectonic information using proxy indicators such as relative sea level changes in tectonically active areas, and archaeoseismology. This is because actual and expected surface observations associated with fault rupture will vary depending upon the fault behaviour model specified in the deformation modelling procedure, which, as noted above, has a direct influence on the way in which a fault grows, the way in which displacement changes along faults, and the way in which slip accrues over time.

## 4.3 Geology[20] (Figure 4.7)

This subsection aims to provide a general outline of the geology of the study area, and has been compiled from field mapping during 1996, 1997/early 1998; existing stratigraphic research by Rondogianni-Tsiambaou (1984); existing publications, for example the 1:50,000 geological map published by IGME (1965) especially for information on the ages and delineations of the pre-rift limestones (i.e before the onset of late Miocene rifting). A more detailed review of the syn-rift (i.e. after the onset of Miocene rifting) stratigraphy, particularly the extensive literature by French workers, can be found in Rondogianni-Tsiambaou (1984) summary in and (Ganas 1997).

### 4.3.1 Pre-Rift Stratigraphy

The pre-rift stratigraphy of the field area comprises predominantly carbonate rocks with an ophiolitic suite emplaced during the Late Jurassic - Early Cretaceous

---

[19] Recurrence is defined by Schwartz (1988) as the frequency of occurrence of earthquakes of various magnitudes, up to a maximum, on a fault or in a region.

[20] The dip and strike measurements given in this section were taken by the author during the field mapping and the single quoted measurement is representative of localised vicinity, for example the limestone area of Almyros, or Donkey Island.

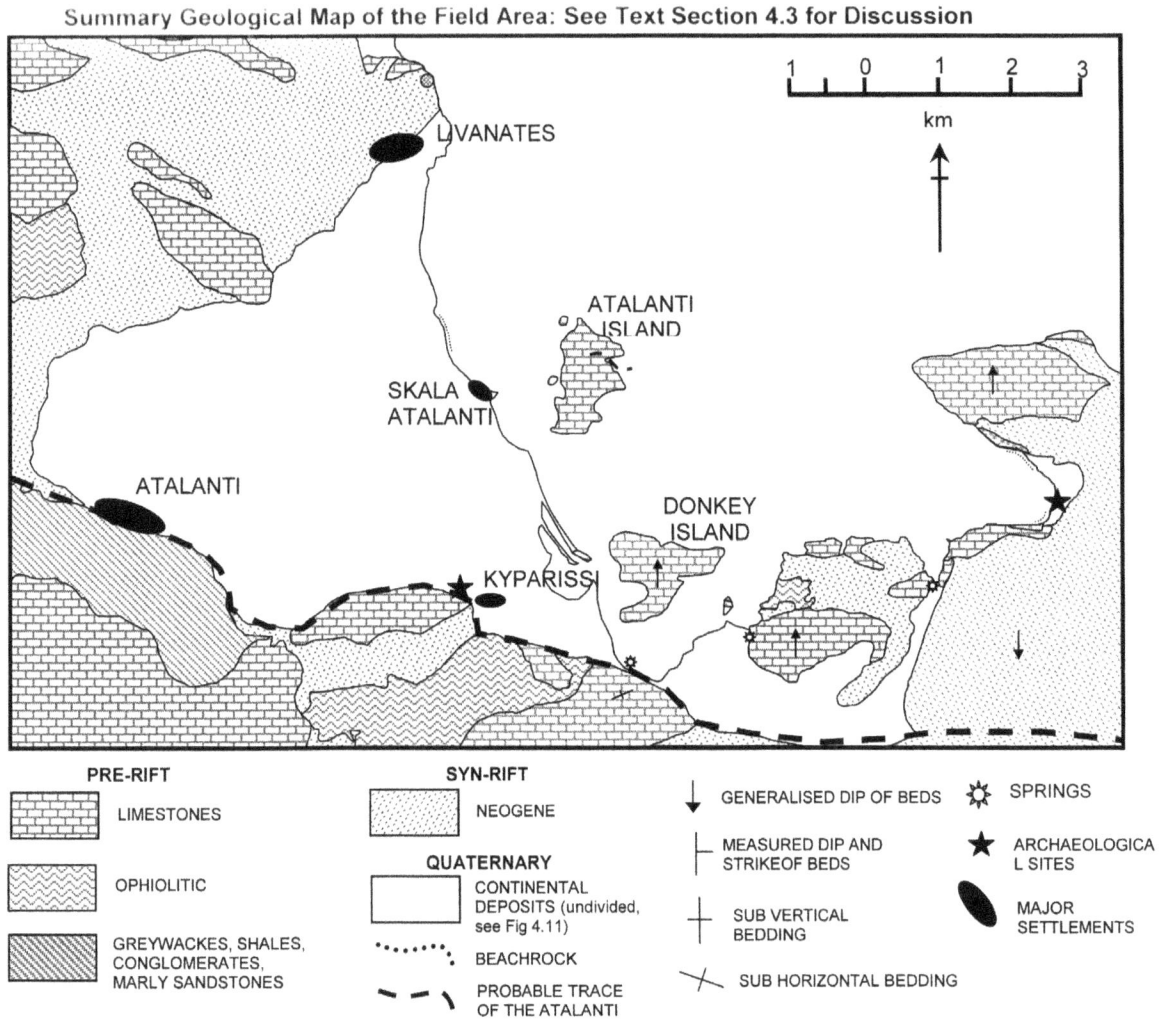

FIGURE 4.8: SUMMARY GEOLOGICAL MAP OF THE FIELD AREA, ATALANTI PLAIN, CENTRAL MAINLAND GREECE.

The pre-rift stratigraphy crops out in the footwall of the Atalanti fault with Permo-Carboniferous greywackes, conglomerates, shales and marly sandstones in the locality of the town of Atalanti. Moving eastwards along the footwall of the Atalanti fault these give way to Jurassic limestones with some Triassic dolomites. Above the village of Kyparissi in the lower slopes of the Khlomon range, ophiolites crop out, passing into Jurassic limestones with sub vertical bedding in the immediate vicinity of Almyros. It is worth noting that these bedding planes may have been previously misinterpreted as the remnants of the surface ruptures associated with the 1894 Atalanti earthquake. Although the bedding planes are clearly not seismic features (no evidence of any lineation have been found on any of the planes examined on the lower slopes) it is likely that the bedding planes constitute a pre-existing plane of weakness that would probably have been re-activated during the earthquake.

East of the village of Tragana, the upland plateau between Proskynas and Malesina comprises almost exclusively back-tilted (068°\16°S) Neogene syn-rift deposits, except for the most northerly point of the

Malesina peninsular (north of the town of Theologos) which is primarily composed of Jurassic/Triassic limestones, dipping to the north (090°\50°N) and heavily intersected by WNW trending travertine fissures. Similarly, the small topographic high to the west of Vivos comprises mainly back-tilted Neogene deposits, though Jurassic limestones (strike between 80° and 90° with a dip ranging from subvertical to observable north to north east [i.e. around 60°] dipping north) crop out on both sides of the bay. Small igneous outcrops are present locally (IGME 1965), especially in the area of the strings at Tragana and at the 20m altitude on the Kavkalo Hill.

Both the islands in the bay (Atalanti and Donkey), and the small bedrock outcrop on the end of the outer spit are predominantly upper middle Triassic limestones dipping south (150°\38°S). The Triassic limestone outcrop (098°\56°S) on the outer spit contains old travertine fissures, again trending WNW which are similar in trend and composition to those on the Malesina peninsular. West of Atalanti, the upland plateau of Xerovouni is predominantly syn-rift Neogene deposits with small irregular outcrops of Jurassic limestones, Triassic dolomites and igneous rocks (IGME 1965). The Neogene on this plateau is not back-tilted, and in some areas appears to be almost horizontal and may possibly indicate its location in a transfer zone or within a persistent segment boundary. The lack of large exposures (this area is well cultivated with olive trees) and observations of true dip (dip angles of less that 5°) made it difficult to place any confidence in any dip and strike measurements taken.

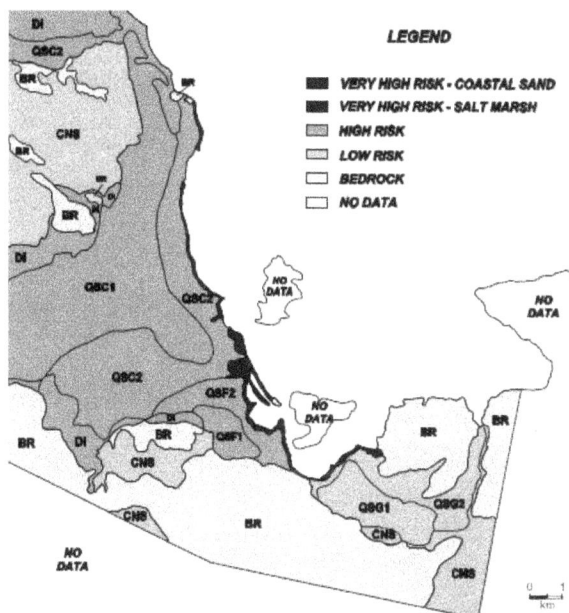

FIGURE 4.9: A SUMMARY GEOTECHNICAL MAP OF THE ATALANTI PLAIN REDRAWN FROM ANGELIDES (1990). THE SHADING REFLECTS THE DATA PRESENTED IN TABLE 4.1 WITH THE DARKEST COLOURS INDICATING AREAS OF POTENTIALLY HIGH GROUND ACCELERATION.

### 4.3.2 Syn-Rift Stratigraphy[21]

A basal unconformity separates the top of the pre-rift rocks from the base of the first syn-rift deposits. The age of this unconformity is approximately 5 Ma (Rondogianni-Tsiambaou, 1984). These deposits are Pliocene lacustrine sediments, best exposed in the Malesina Basin and are composed primarily of marly limestones with intermittent sandstones.

### 4.3.3 Quaternary

The data for the Quaternary was collected from both direct sources (i.e. field observations and mapping in 1996, 1997 and 1998) which are included in the summary paragraphs and from the published IGME geotechnical report by Angeldes (1992). The relevant data has been tabulated and is presented in Table 4.1 with corresponding map shown in Figure 4. 9). No data were collected for the predominantly pre-rift bedrock Atalanti

and Donkey Island. Small patches of salt marsh and sand deposits can be observed on Donkey Island; however Atalanti Island is in private ownership and permission to visit could not be obtained at the time of the visit.

Pleistocene deposits sit conformably on the Pliocene. Lignites signal a change from a lacustrine to a fluvial (conglomerate) depositional environment, with some lagoonal sediments observed in coastal localities (Theologos) (Rondogianni-Tsiambaou, 1984). By the Mid-Pleistocene (0.5 Ma) these are replaced by continental deposits. The base of the Quaternary is indicated by a marine incursion/transgression, followed by continental deposits such as red soils, alluvial fans and talus breccias. The Holocene deposits (from 10,000 Ky) range from coastal (beach rock) to fluvial sediments, with mud beds and remnant salt marsh in the inter-tidal margin of the coastal plain area. The occurrence of back-tilted Neogene lactustrine deposits in the footwall of the Atalanti fault is indicative of neotectonic activity (Ganas & Buck, 1998). This is corroborated by the presence of dissolution marks and small erosion shelves/bands on the footwall of the Jurassic limestones at the quaternary/bedrock contact in the area of the Almyra springs. Ideally, a detailed review of the geotechnical and geomorphological information of both the archaeological site and the hinterland is required for an archaeoseismological appraisal as this will give some indication of the range of processes which may be operating at the site and which could potentially be the mechanism for any observed damage. These processes can result in both primary and secondary seismic damage (McCalpin & Nelson, 1996) and mechanical surface processes such as non-seismic landslip and slumping.

#### Sediment Types
The field study area can be roughly subdivided into four zones, based on the geographical location and the constituent deposits, of which three are located in the plain area and comprise Holocene unconsolidated sediments. These are the coastal margin; the plain; and the intermediate region (i.e. the between the slope and the plain); and will be discussed in this order below. The fourth zone is the surrounding bedrock hills, and this has been discussed in the preceding section on pre-rift geology.

#### Coastal Margins
The deposits in this zone reflect their close proximity to the coast line, and are primarily composed of fine grained sands, silts and clay with no cohesive qualities, and waterlogged in the case of the silts and clays (Angelides, 1992). The coastline is marked in the south by salt marsh,[22] and to the north the immediate beach area is shingle/pebbly beach.

---

[21] The palaeodepositional environments for the syn-rift suggested by Rondongianni-Tsiambaou (1984) have been used here.

[22] The salt marshes appear to be in a deterioration stage due to the marsh drying out as a consequence of a change from intertidal to terrestrial environment (Cundy pers comm), or possibly as a result of water abstraction associated with the heavy horticultural industry in the area.

**Coastal Margin Deposits**

| Fig 4.11 Shade | Angelides code | Composition | Cohesion | Permeability | Ground Water Table (GWT) | Seismic Risk |
|---|---|---|---|---|---|---|
| | - | Sand | None | VERY | Deposit at @ sea level | VERY HIGH |
| | - | Soil/Silt/Clay | - | Waterlogged (>75% time) | Deposit at @ sea level | VERY HIGH |

**Plain Deposits**

| Fig. 4.11 Shade | Angelides code | Composition | Cohesion | Permeability | Ground Water Table (GWT) | Seismic Risk |
|---|---|---|---|---|---|---|
| | Q.S.F1 | Sandy Material (Up to 3m = sand) | LOW | SEMI | 5 - 10 m | HIGH |
| | Q.S.C1 | Clay & Sand | LOW | SEMI | 5 - 30 m | HIGH |
| | Q.S.G1 | Sand & Gravel | LOW (+ with depth) | VERY | > 10m | LOW |
| | Q.S.F2 | Sandy | - | VERY | @ 5m | HIGH |
| | Q.S.C2 | Clay & Sand | (cont. of line 2) | LOW | - | HIGH |
| | Q.S.G2 | Sand & Gravel | - | VERY | 5 - 20m | LOW |

**Intermediate Deposits (i.e. between the slope and plain)**

| Fig. 4.11 Shade | Angelides code | Composition | Cohesion | Permeability | Ground Water Table (GWT) | Seismic Risk |
|---|---|---|---|---|---|---|
| | Dl | Coarse grain Sand & Gravel | LOW | HIGH | - | HIGH |
| | C N S | Clay, Marl, Sandy Clay | - | - | - | LOW |

**Rocky Formations**

| Fig. 4.11 Shade | Angelides code | Rock Type | Cohesion | Permeable | Ground Water Table (GWT) | Seismic Risk |
|---|---|---|---|---|---|---|
| | L.N | Marly Limestone | - | YES | >30m | LOW |
| | L.j | Jurassic Limestone | - | VERY | 15 - 65m | LOW |
| | V.R | Ophiolites | - | - | Not Measured | LOW |
| | S.c.h.v | Schists & Cherts | - | - | Not Measured | LOW |
| | P.L | Palaeozoics (Volcanics) | - | - | Not measured | LOW |

**Table 4.1:** Geotechnical properties of the sediments in the Atalanti plain derived from Angelides (1992). The shading in the first column correlates with the shading on the map at Figure 4.9.

*Plain Deposits*

The sediments in the Atalanti plain comprise predominantly low cohesive (increasing with depth) sand, and gravels with intermittent clays. The ground water table is between 5-10m in the main plain region (Angelides, 1992), rising to up 20m in the Mikro Vivos and Vivos bays and associated valleys (observed and noted during well drilling activities associated with farming in the area, during the field mapping in the summer of 1997). The variable mixture of sand, clay and gravels renders the Atalanti plain highly suitable for agriculture, specifically horticulture in the Almyra region, where freshwater springs rise from the quaternary/bedrock contact, see. Offshore of the Almyra area, between the mainland and Donkey Island, the near-shore sediments are largely clayey. These grade to sand

and sandy clay to the north (towards Atalanti Island) and loose rocks with clay mud matrix to the south (noted during field observation in 1996/97 and 1998).

*Intermediate Zone*
The intermediate zone, i.e. between the sub-horizontal plain and the bedrock mountains (i.e. the Xerovouni plateaux), comprises mainly low cohesive, highly permeable scree deposits with large outcrops of Neogene (see syn-rift section 4.3.2. above). The scree, or talus, is dominantly limestone and ophiolitic material from the pre-rift limestones.

*Data Confidence*
A primary study of the quaternary sediment has not been carried out to verify or 'ground truth' the figures given in Angelides report as this requires highly specialised techniques. However, for the purposes of this appraisal it is reasonable to accept the data as the general picture that Angelides presents correlates with that known from post earthquake research; that the shaking and liquefaction occurs more readily in materials with the least cohesion, caused by factors such as high water content and low cohesive values (Reiter, 1990). Additionally, the work published by Angelides (1990) is within his sphere of expertise and given the interdisciplinary nature of this work it is reasonable to assume that certain data sets would need to be obtained from workers within the fields of specialism.

*Ground Truthing of Existing Geological Data*
As noted in the introduction to the geology subsection, the data presented here are a combination of previous publications and personal field mapping and observations. Although it was not possible to check the divisions of the pre-rift lithology (which is primarily based upon fossil evidence derived from sites across Greece), it was immediately apparent that the structural information contained on the IGME geological map (Atalanti sheet) was on a number of occasions misleading, if not incorrect. Therefore, only primary observations of structural data – specifically measurements – have been used in the construction of the geological map shown in Figure 4.8.

Obviously, there is always some variation in the readings of orientations (azimuths) due to change in magnetic north over time (IGME mapped the Atalanti sheet in 1961 and 1963) as well as variations in the instruments and methods of measurement. Therefore, apart from the simplicity of the published map in terms of lack of detailed lithological subdivisions, only major contradictions are listed below:

1.  the Triassic limestones on Atalanti and Donkey island dip to the south and not to the north, and
2.  the stratigraphic column does not contain the basal conglomerate that is indicative of the switch from pre-rift to a syn-rift depositional environment.

### 4.3.4 Topography

The field study area can again be divided into three general zones, based on slope angle, which roughly correlate with the bedrock geology and the deposits noted in the preceding section. The first zone is the mountainous area in the footwall of the Atalanti fault, including the forested area behind the town of Atalanti, and the bedrock outcrops in the district of Almyra and Kyparissi, specifically Gardinitsa and Palaiokastro. Slope angles of over 30° were noted by Angelides (1992), but outcrops of shear, subvertical (jointed) bedding was also observed during field mapping.

The second zone consists of slope angles of between 10° - 30°, and primarily includes all the areas with ophiolitic and pre-rift limestones (other than the subvertical bedded, i.e. the intermediate zones). This includes the immediate fault footwall region behind the archaeological site at Kyparissi, and the Xerovouni plateau. Much of the lower sections of the intermediate zone which correlate to the gentler slope angles have been artificially terraced for olive production.

Sub-horizontal slopes, or slopes of less than 10°, make up the third zone, and are predominantly the plain localities, up to the 20m contour. This area also includes the relatively flat low-lying valleys associated with the small inlet bays such as Theologos, Vivos, and Mikro Vivos. Again, accurate measurement of these areas is hampered by extensive agricultural activity, though the pre-intensive agricultural elevations would not have resulted in these areas being included in the second, intermediate, zone.

### 4.4 Regional Seismicity

The range of evidence for seismic activity in a given region spans the human (i.e. contemporary and derived written accounts) and the physical (i.e. instrumental) sources. This primarily reflects the relative youth of instrumental records (the establishment of the permanent seismological network took place in 1965) and the longevity of the written record in Greece. This section provides layers of instrumental data ranging from the modern record, via textual accounts from the contemporary records of the 1894 earthquake sequence, to literary accounts from ancient texts. No geological evidence of earthquakes on the Atalanti fault is included, due to the fact that there are only two probable exposures. The first, in cross-section, occurs in a disused quarry behind the town of Tragana. The second, is an along strike exhumed mirror plane at the rubbish dump in the region of Asprorevma. Although there was up to 1 metre of displacement associated with the 1894 earthquake (Skouphos, 1894), the ground ruptures cut Quaternary deposits and have not been securely identified in the field so as to allow any consideration of this geological evidence.

## 4.4.1 Instrumental Data

Instrumental data on earthquakes within the field area are available from July 1964 to the point of request (November 1996) from the National Observatory in Athens. The data specify the calendar date, time, co-ordinates in latitude and longitude (in decimal degrees) and magnitude. All these events are shallow focus events, and the depth is recorded automatically as 5km where better precision cannot be achieved. Data were requested for all events > Ms 3 for the area between Lat. 38.20 - 38.50 and Long. 22.30 - 23.30 and were provided in ASCII format (see Appendix 1). Due to the lack of primary and secondary geologic evidence of earthquakes below moment magnitude 5 and 4 respectively (McCalpin & Nelson 1996), the data were processed to remove smaller events, along with those with foci occurring outside the refined field area limits of Lat 38°20' - 38°50' and Long. 22°30' - 23°30'. The remaining events are presented in the table at Appendix 2 and in the image generated using IDRISI V.1.01.002 Geographical Information System (GIS).[23]

## 4.4.2 Contemporary Historical Sources

This section discusses accounts written contemporaneously or very soon after seismic events. For the distant past this will include epigraphic evidence, whereas for the recent past newspaper reports, papers in scientific journals and descriptive articles in special editions are the main sources of data. Due to the size of the tables, these have been attached as an appendix.

With the exception of the scientific journals, the authorship of most of the newspaper accounts reviewed is anonymous and the details of information contained within these accounts must be viewed with caution. It is possible that effects of the earthquake have been exaggerated either to provide dramatic impact for the reader or because of the use of indirect sources. Notwithstanding, the articles can provide a valuable, albeit rough, guide to the type of damage and the geographical extent of the effects of the earthquake. Similarly, the descriptive accounts in scientific journals may provide details of ephemeral phenomena which, if visible today, may corroborate a new model or hypothesis.

Locating contemporary articles that may include information on the effects of an earthquake is a time consuming task if a 'ball park' date of an event, or at least the knowledge that one has occurred, is not known. Additionally, when cross-correlating the dates of earthquakes it should be borne in mind that different calendars may have been used and thus could lead to a difference in the event date. For example, the first shock of the 1894 Lokris sequence is dated 20 April in the English papers, but, in the Greek paper, ΑΚΡΟΠΟΛΙΣ

(Acropolis) the first mention of the earthquake occurs on 15 April 1894. Ambraseys & Jackson (1990) note that this event came without warning; therefore the difference in reported dates can only be attributed to the use of different calendars, indeed Greece did not switch to the Gregorian calendar until the 1920's. Without knowing that there was definitely an event in April 1894, it is would have been easy to miss this report or unwittingly conclude that more than one event occurred at the time.

Although there is, at present, no epigraphic evidence of the earlier seismic events in the Atalanti region and little accessible contemporary literature pertaining to the Byzantine or Ottoman period (Ambraseys & Jackson, 1990), there is considerable contemporary literature, both in the popular press and scientific treatise, for the 1894 events. Certainly, for the popular accounts, the geological and geographical details are to be found within a larger descriptive text; for ease of reference this has been tabulated and can be found at Appendix 4.

At present this table contains primarily English, Greek, French and German references which have been brought to light through the course of this research. Greek reports written in Katharavousa (formal Greek) have not been directly translated and are included here for future reference. The contemporary accounts have been the subject of research by Ambraseys and Jackson (1990) and the information on the extent of damage for the 1894 event is taken from this work. It should be noted that this is not a definitive list of contemporary references to earthquakes in the Lokris area, in as much as it relies heavily, as does any historical or archive research, on the quality of the archive, the relative ease of access to documents once located, and one's ability to read the language of publication. However, it does provide many details of the ephemeral surface phenomena, which are important considerations in any archaeoseismological appraisal.

Changes in regional landscapes occur over time due to numerous contributory and often barely perceptible factors. In the Atalanti plain this is no more apparent than in the vicinity of the major civil works associated with the upgrade of the National Road (E75). Across the plain, small localised changes can be identified. Perhaps the most obvious is the blocking of the natural drainage of the plain by the civil works, causing a seasonal build up of standing water on the landward side of the carriageway. However, changes are not only the result of human influence but also of natural phenomena, for example earthquakes. For seismic events occurring in the recent past, these changes can be measured, and to a certain extent the permanence of change quantified, by simply using 'before and after' photographs. Similarly, a comparative style analysis of written observational data can also be used for the more distant, pre-photographic, events where there is a long written historical record. However, teasing out brief topographical observations from historical texts requires previous knowledge of the

---

[23] See Chapter 5 Geographical Information Systems

date of a seismic event, and more importantly detailed objective reading of contemporary historical and, in the Mediterranean region, classical literature. Knowledge of historiography style is essential in these cases.

In summary, contemporary accounts of earthquakes can be used as a general indicator of the possible ephemeral seismic phenomena which may have occurred at a particular archaeological site, and which without these accounts would be totally unknown to the appraisal. For example, the extent of landslides and or ground slumping associated with the earthquake in cases where there is now no surface evidence for these phenomena and for which villages, or even which sites, may have been affected. The importance of such information is simply that deformation or damage observed in a site would, without these observations, have been attributed as direct seismic damage rather than secondary surface seismic effects.

### 4.4.3 Literature Concerning Physiographic Changes in the Landscapes

Topographic studies carried out by classicists in the late 19th century can similarly provide useful observations on the natural environment. For example, one of the most detailed accounts of local topographies can be found in Frazer's (1913) accompanying notes to Pausanias' *Guide to Greece*. However, research that specifically seeks to confirm a classical text should be treated with caution, as it may be written with an objective other than to provide a purely descriptive guide; and this should be taken into account when using any derived information in further research. Ambraseys (1996) emphasises the importance of considering the context of literature when using historical sources in palaeoseismological studies.

*Travellers*
The existence of monumental architecture and classical texts has ensured that Greece has always been an attractive proposition to the traveller, many of whom attempted to identify and locate geographically the cities and temples of the classical texts (Frazer, 1913). The early 'travellers', such as Gell (1819), made accounts of their journeys, often with an accompanying map, and often published their work as guides for use by others wishing also to visit the area. Due to the highly descriptive nature of their texts and their lavish publication, many of these volumes can provide pre-photography topographic details. Similarly, the intricate and detailed plates and line drawings that accompany the texts can be used to identify topographic and environmental change.

Unlike historical literary accounts that may use references to seismic activity or the occurrence of natural phenomena (e.g. eclipses) within a biased narrative, these travel accounts seek to gain nothing by describing to the readers of the 18th and 19th century the physical landscape, for example the proximity of villages to rivers and coasts and the prevailing vegetation. Similarly, the

inclusion of geological phenomena such as earthquakes or volcanic activity, with then largely unknown mechanisms, are included for no other purpose than to enhance the descriptive narrative. (See Guidoboni *et al.* (1989) and Ambraseys and White (1996) for a thorough discussion and bibliography on the use and interpretation of historical documents for authentication of geological phenomena). A modern analogy of such accounts would be the *Rough Guide* or *Blue Guides*, which have their origin in the early '*Baedekker*' guides to travelling published in the late 19th century.

Often journeys are organised within the narrative as a number of excursions, each contained within a small chapter, describing what was seen *en route* with a map showing principal sites of habitation and including names and geographical or topographical information such as hills and rivers. Occasionally accounts written as travel companions are accompanied by the time it would take to travel, by a specific mode, between certain points on the journey. For example, Gell (1819) not only records azimuths and measurements in his original notebooks, but gives the travel times by foot. In these accounts natural topographic phenomena, such as the location of ravines and riverbeds, serve as geographical points of reference, and it is reasonable to assume that they have been accurately described as the sole object of the text is to guide the subsequent traveller.

After an extensive literature search of the early travellers in Greece, specifically the central mainland regions, it was found that very few had visited, or even passed through, the region of Opuntian Lokris. It would appear that the more central routes to the North, through Thiva, Lamia and Larissa towards Thessaloniki were preferred, probably due to the trail of visible relics in the region of Boiotia, such as Orchomenos, Gla, and Chaironea. Those that did visit, and who have provided written accounts, were primarily concerned with the identification of the Homeric city of Opus, and their observations are made in relation to this question.

Perhaps the most important topographic inference that can be drawn from the modern historical literature is that there has been significant change in the coastline in the Atalanti area, with some apparent loss of salt marsh, due to subsidence during the 1894 earthquake. This is supported by the fact that the once significant late 19th century salt production industry, which was present in the Almyra area, was abandoned when the salinas were flooded after the 1894 earthquake sequence. Table A3.3 at Appendix, 3 gives a summary of the pertinent geographical information found in the itineraries.

*Historical Maps*
Cartographic data can be used in the same way as the comparison of photographs for pre- and post-earthquake topographical changes. However, this is only possible where an observable change in a specific geographical location has taken place, for example the disappearance

| YEAR | DATE | LAT | LON | Ms | INT | REGION |
|------|------|-----|-----|-----|-----|--------|
| 426 BC | WINTER | 38.5 | 23.1 | 6.6 | VIII | Orchomenos (Thucydides) |
| 426 BC | SUMMER | 38.8 | 22.6 | 7.0 | IX | Phthiotis, Scarfia (Strabo), Lokris (Thucydides, Diod. Sic.) |
| 348 BC | - | 38.4 | 22.5 | 6.7 | VII | Delphi |
| 279 BC | - | 38.4 | 22.6 | 6.8 | IX | Delphi |
| AD 106 | - | | | | VII | Oreoi, Evvia and Opus |
| 551 | - | 38.9 | 22.7 | 7.0 | X | Phthiotis, Achinos |
| 1321 | - | 38.3 | 23.3 | 6.3 | VIII | Thebes |
| 1421 | Sept 18-22 | | | | | Evvia (Stiros 1985) |
| 1544 | 22 April | 38.8 | 22.6 | 6.8 | IX | Lamia |
| 1694 | July | | | | | Chalkis (Stiros 1985) |
| 1740 | 22 July | 38.8 | 22.6 | 6.5 | VIII | Lamia |
| 1758 | May | 38.9 | 22.7 | 6.8 | | Lamia, Lichades (Stiros 1985) |
| 1853 | 18 August 08:30 | 38.3 | 23.3 | 6.6 | X | Thebes, Boiotia |
| 1870 | 1 August 00:41 | 38.5 | 22.5 | 6.8 | IX | Arachova, Boiotia |
| 1874 | 18 March 05:00 | 38.5 | 23.5 | 6.0 | | Eretria or off shore Anthedon (Stiros 1985) |
| 1893 | 23 May 22:02 | 38.3 | 23.4 | 6.2 | VIII | Thebes |

**Table 4.2**: Historical Seismicity in the Lokris, Boiotia, and Lamia areas of central Greece. The 426 BC event is included here as not all authors agree that the Atalanti fault is the seismogenic source of this event, for example Bousquet & Pechoux (1977) note the epicentre in the Sperchios Basin offshore from Lamia. The principal references are Ambraseys & White 1996, Ganas 1997, Papazachos & Papazachou 1997 and Guidoboni 1989, Guidoboni *et al* 1989. Other sources are indicated.

or apparent geographical shift of either a natural or a man-made landscape feature. Such comparative measurements usually only give a general idea of the topographic change, as the empirical data sets, i.e. themaps, are prone to inaccuracy, especially with age. Inaccuracy is due to variables in quality of mapping and the accumulation of errors in map production, both of which are related to the origin and date of construction of the map. Similarly, the accuracy of the map dictates the accuracy of any measurements of the change, always resulting in errors equal to or greater than the known error on the most inaccurate map. In the case of the comparative analysis of old maps with more recent, or even current, sheets one can only derive a general idea, or feel, of the change, and not precise figures, as the errors on the older map will dictate the accuracy of the comparison.

Several pre-1894 maps exist, including the geological maps of the ground ruptures of the 1894 earthquake which are shown superimposed on the pre-event topography (Skouphos, 1894; Philipson, 1951). However, perhaps the most accurate for the purposes of topographic comparison is the 1890 British Admiralty Map and its accompanying book of notes which provides details of the projection, spheroid and errors (The Mediterranean Pilot, 1918). This map clearly shows that prior to the 1894 earthquake Donkey Island was attached to the main coastline by a marshy area. Additionally, the map shows the location of archaeological remains that are no longer visible on the ground namely ancient walls and embankments with the site of Opus located in the Almyra and Kyparissi area of the plain.

### 4.4.4 Ancient Literary Sources

The primary sources of information on the ancient earthquakes reviewed in this section are derived from the classical source book *Paulys Real-Encyclopaedie* (1900), as cited by Guidoboni *et al.* (1989), Papazachos and Papazachou (1997), and Ambraseys and White (1996). All the prehistoric earthquakes that are known to have occurred on seismogenic sources within the field area are included. However, the historical seismic record is imperfectly known, being increasingly incomplete farther back in time. Many other seismic events, not necessarily on the known active faults in the study area, may have caused ground motion in the area but for various reasons remain unlisted. Only the events assumed to have occurred on seismogenic sources within the field area are discussed in detail. Historical events occurring on sources outside the area, but assumed to have caused significant ground acceleration (shaking) in the Atalanti plain, are included in Table 4.2.

The study area loosely correlates with the ancient territory of Opuntian Lokris (East Lokris) and, as previously noted, focuses on the Atalanti plain. There is still much debate on the location of the ancient city of Opus, capital of the region (Fossey, 1990; Dakoronia, 1993), which has throughout the period of archaeological investigation been hindered by the lack of surface architectural remains. Consequently, there is a paucity of references in classical literature and early travel accounts. Pausanias mentions Opuntian Lokris almost in passing (Levi, 1971), and a large volume of literature has been published in the debate concerning the authenticity of the personal observations described in the *Description of*

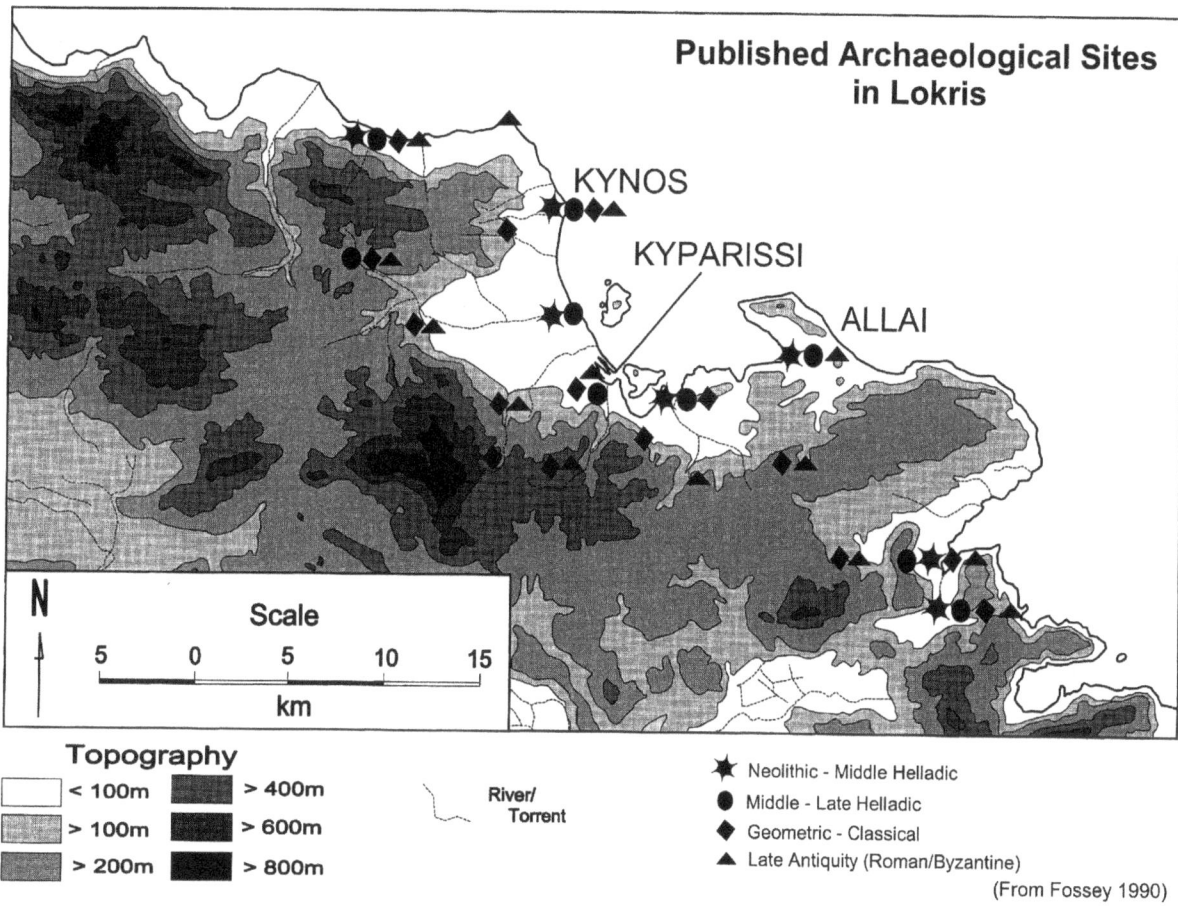

FIGURE 4.10: DISTRIBUTION OF KNOWN ARCHAEOLOGICAL SITES BY PERIOD IN OPUNTIAN LOKRIS (WHICH INCLUDES THE ATALANTI PLAIN) REDRAWN AFTER FOSSEY, 1990.

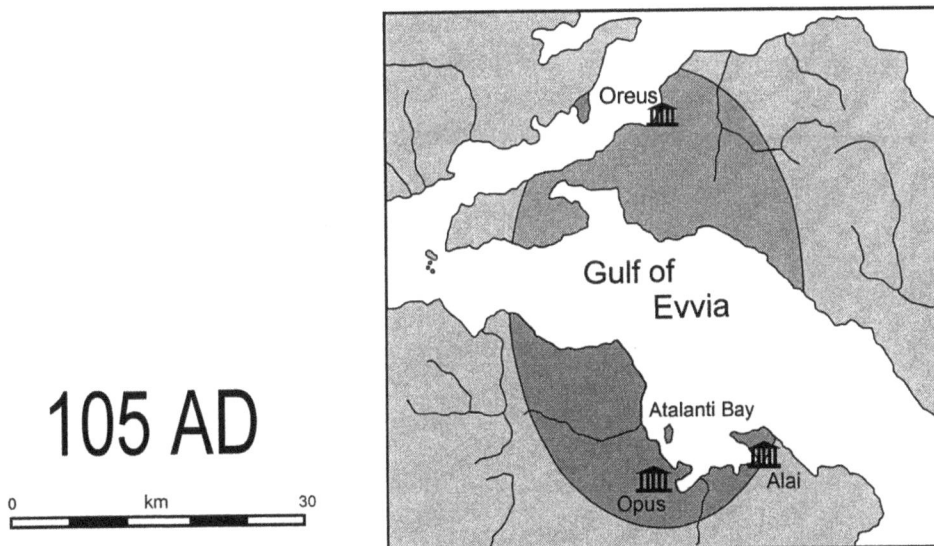

FIGURE 4.11 : THE AREA AFFECTED BY THE 105/6 AD EARTHQUAKE COMPILED FROM HISTORICAL TEXTS/SOURCES. REDRAWN FROM GUIDOBONI ET AL., (1989).

*Kynos*

| Earthquake Date | References | Evidence Cited |
|---|---|---|
| Mid 12th C BC | Dakoronia, 1996 | Lateral shift in mudbrick walls, dislocations of mudbrick walls, mudbricks that had fallen into pithoi [giant terracotta jars] p 41. |
| Mid 12th C BC (<100 yrs after) | Dakoronia, 1996 | Various dislocations (unspecified), conflagration, abundance of marine fossils and rounded pebbles. p 42 |

**Table 4.3**: Summary of the evidence cited for the presence of earthquake damage at the site of Kynos.

*Kyparissi*

| Earthquake Date | References | Evidence Cited |
|---|---|---|
| 540 BC | (Stiros 1985) | Pottery evidence is used to suggest that there was an earthquake between 540 – 500 BC, suggests that the stoa was destroyed and abandoned after 540 BC but before 500 BC |
| 426 BC | (Stiros & Dakoronia 1989) | Deformed and postulated fracture of architectural elements of a stoa. |
| 520 BC | (Stiros 1988a) | The observed deformation, and postulated fracture of an stoa wall constructed of 1m long ashlar blocks. |
| 480 BC | (Dakoronia 1993) | Destruction indicated by the dates on the tiles used for roof repairs. No specific evidence of earthquake cited only that earthquake is feasible due to its close proximity to the Atalanti fault. |
| 450 BC | (Stiros & Pirazzoli 1995) | The damage of the architectural remains of the stoa at Kyparissi. |

**Table 4.4**: Summary of the evidence cited for the presence of earthquake damage at the site of Kyparissi

*Alai*

| Earthquake Date | References | Evidence Cited |
|---|---|---|
| 510 BC | (Goldman, 1940) | A building was razed and buried under a thick covering of earth sometime after 510BC |
| 426 BC/ 425 BC | (Goldman, 1940) | Destruction of the first temple area attributed to earthquakes through a comparison of the architectural terracotta of new buildings with those at the Argive Heraion by Prof. Dinsmoor. |
| 480 BC | (Wren, 1996) | Destruction of the first temple and the apparent lack of continuity of occupation of the site, identified through the lack of classical material. |

**Table 4.5**: Summary of the evidence cited for the presence of earthquake damage at the site of Alai

*Atalanti Island Fortress*

| Earthquake Date | References | Evidence Cited |
|---|---|---|
| 426 BC | (Lolling 1876) | Classical references discussed in terms of the physiographic setting of the islands. No actual information on any observed destructions in the fort area. |

**Table 4.6**: Summary of the evidence cited for the presence of earthquake damage at the site of the Atalanti Fortress on Atalanti Island

*Greece*. Early commentaries upon this work suggest that he did not actually travel there himself but refers to earlier texts by authors such as Polemon of Illion (Habicht, 1985). Only brief mention is made of the Locrians (the name by which the inhabitants of the region are known) by Homer in his '*Catalogue of Ships*' (Hope-Simpson and Lazenby, 1970) and as the catalogue is an entirely different writing context it provides little topographic detail other than that there was a sea port in the Atalanti plain (at Skala Atalanti, previously called Kato Pella).

*427/426 BC*

The 426 BC event is known from its inclusion in three prominent and widely available classical texts. The first, and perhaps the most reliable, is the contemporary historical account of the Peloponnesian Wars by Thucydides which provides the relative date of the event (*Thuc. iii. 89; Loeb,ii.157-9*) (Lolling, 1876; Mouyiaris, 1988). The second two are derived accounts, one from the historical library of Diodorus of Sicily (*Diod. Sic.,xii.1ix/Loeb, v. 49*) and one from the geographical account of Greece by Strabo (*Strabo, I. iii. 20/Loeb, ii.223-7*), both of which were written almost a century after the actual event. Many authors, making direct comparisons in the surface effects of the 1894 Atalanti event and the destruction described in the three classical texts, consider that the Atalanti fault was also the seismogenic source of the 426 BC earthquake (Rondogianni-Tsiambaou, 1984; Stiros & Rondogianni, 1985Mouyiaris, 1988; Stiros, 1988; Ambraseys & Jackson, 1990; Ganas, 1997). Translations of the Greek texts by Professor Graham Shipley can be found at Appendix 4.

*AD 105/6*

There is only one vague classical reference to this event in the chronicle of Eusebius (early 4th century AD). Guidoboni *et al.* (1989) considered this to be a reliable source, and on the strength of this the event is included in the recent catalogue of earthquakes in Greece (Papazachos & Papazachou, 1997). Opus is specifically noted as having been destroyed by the effects of the earthquake. Apart from listing the city name, no other details are given. The event is variously dated AD 107, 106 or 105 depending upon which translation of Eusebius is used; see Guidoboni *et al.* (1989) for a full list of subsequent catalogues that have included this event. In *Pauly Wissowa Supp.V 2*, Oldfather (1890) includes a reference to the event with regards the possible relocation of the city of Opus, but this is pure conjecture, as there does not appear to be any supporting evidence for this statement.

It would be foolish to try to assign the exact source of this event based on such scant references. However, from the known information and the map included in Guidoboni *et al.*, (1989) (see Figure 4.11) it is not unreasonable to consider that this event may have been the result of an earthquake resulting from rupture on one of the known faults in the field area, i.e. the Atalanti Fault, the imprecisely known offshore (Theologos) fault, or any one of the smaller intra-basinal faults in the Gulf identified by (Smith *et al.*, in prep.)

*Events on Seismogenic Sources Outside the Field Area*

The events in Table 4.2 are derived from existing catalogues and include events which are known to have occurred in the Thiva basin and the Lamia/Sperchios graben. The significant distance of the sources from the field area (as much as 40km away) means they do not warrant a detailed discussion, but they are included here because of the possibility that appreciable ground motion may have occurred in the Atalanti plain, and as such, these events should also be considered as possible contributors to any observed archaeoseismic damage. Cautious use should be made of this composite list, especially in terms of epicentral location and earthquake magnitude, which are often derived from generalised accounts of earthquake effects and observed damages

## 4.5 Archaeology of the Atalanti Plain

Although there are few above surface archaeological remains, the Atalanti Plain (and environs) is now known to be rich in a broad spectrum of archaeological material from the Neolithic to the Byzantine Period (see Figure. 4.10.). *The Ancient Topography of Opuntian Lokris* by Fossey (1990) gives the most comprehensive account of the existing literature, on a site-by-site basis, throughout the archaeological periods (Hellenistic, Roman, etc.), and formed the starting point from which the archaeological research for this research took place. In addition to the published literature of supposed seismic damage, the present author made a number of field observations during the course of the fieldwork that were indicative of seismic activity, for example submerged remains. However, it has not been possible to include these observations due to the restrictions of the archaeological permit with regard to previously unpublished information.

### 4.5.1 Previously Cited Archaeological Evidence for Earthquakes

In this subsection, information relating to previously published observations of seismic damage in archaeological sites within the field area is presented in an impartial tabulated format, giving the assigned date of the event (based on the archaeological evidence), the published reference, and a brief summary of the cited observations from which a seismic mechanism of destruction was derived. A more detailed discussion of the information can be found in Buck and Stewart (2000). It should be noted that the paper does not include a discussion of the Atalanti fortress archaeological site because the reference was discovered and translated after the acceptance of the original manuscript.

## 4.6 Earthquake Effects on Archaeological Sites

There are two principal effects of earthquakes on archaeological sites which need to be considered in an archaeoseismological research. These are (1) deformation (or the permanent change in surface topography), and (2) ground acceleration (commonly referred to as ground motion or shaking).

### 4.6.1 Deformation Models (Vertical Displacement)

Mathematically derived models of ground dislocation associated with normal faulting have shown that there is not only a higher ratio of co-seismic uplift to hanging wall subsidence, but also a differentiation in the expected displacement, with maximum subsidence and uplift occurring in the area immediately adjacent to the fault plane and decreasing to zero with distance away from the initial rupture (Ma and Kusznir, 1992). Further, it is shown that displacement is variable along the strike in relation to the fault behaviour pattern; i.e., for characteristic behaviour, the maximum displacement is found in the centre of the segment, decreasing towards the segment boundaries. These mathematically predicted phenomena have been corroborated by interferometric observations of surface displacement on normal faulting earthquakes, measured using orbital radar system (SAR) techniques, for example the 1995 Grevena (northern Greece) earthquake Meyer *et al.* (1996).

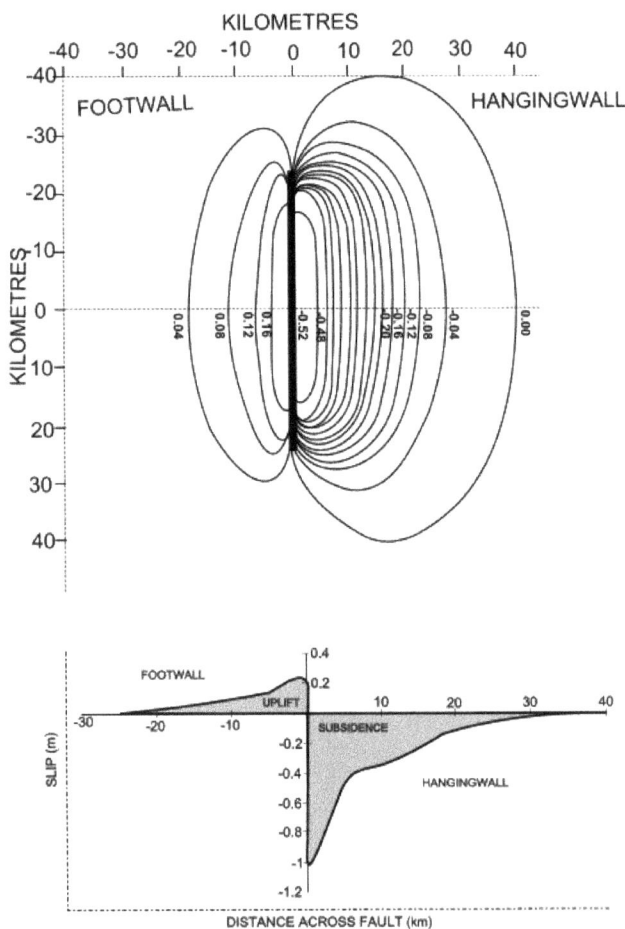

**FIGURE 4.12:** DEFORMATION ELLIPSES (REDRAWN FROM MA & KUSZNIR (1995)) AND ASSOCIATED DEFORMATION CURVE (REDRAWN FROM GANAS, 1998)

### 4.6.2 Ground Motion

The principal factors affecting seismic induced ground motion are earthquake magnitude and seismic wave attenuation (Reiter, 1990).[24] Given the mathematical formulas for and between impedance (resistance to rock or soil particle motion) and absorption (dampening or anelastic attenuation), and excluding all other contributory factors, the amplitudes of seismic waves of equivalent distance from an earthquake epicentre will generally be higher on low density (low velocity) soils and sediment and lower on high density (high velocity) rock (Reiter, 1990). Furthermore, in general there is a decrease in wave amplitude and an increase in wavelength away from the source. However, recent fieldwork following large magnitude earthquakes has shown that observed structural/architectural damage from ground motion is more dependent upon local-near surface conditions, often referred to as 'site effects', which include the properties and configuration of the material (both natural and anthropogenic) through which the seismic waves propagate (Reiter, 1990; Brumbaugh, 1999;). Different combinations of velocity and density properties in the surface deposits will produce varying responses to ground motion that can in turn be extremely localised depending upon the configuration of the surface deposits. This is highlighted in the damage maps produced after large magnitude earthquakes such as those for the 1906 San Fancisco and 1989 Loma Prieta earthquakes (Yeats *et al.*, 1997).

*Seismic Risk 'V's Seismic Hazard*

The terms seismic 'hazard' and 'risk' are generally perceived by non-specialists to mean roughly the same thing and are often used interchangeably: however, they are not the same. Seismic hazard is the physical phenomenon that underlies the danger (i.e. the earthquake and resulting ground shaking etc.), whereas seismic risk is the likely human and/or economic cost that results from the hazard (Reiter, 1990; Yeats *et al.*, 1997). This relationship is described in Papazachos and Papazachou (1997) by the equation:

$$R = H*V$$

Where $V$ is the vulnerability of the structure (based on the technical aspects such as materials and quality), $H$ is seismic hazard, and $R$ is seismic risk.

In tectonically active areas such as the Mediterranean regions, seismic risk is becoming increasingly more important within archaeology, especially in the field of cultural resource management (Nikonov, 1988; Stiros, 1988b). This is due, in part, to the existence of a high number of above-ground architectural remains, for example temples, fortresses and churches, and the presence of large settlement sites such as the so called palaces of Knossos in Crete and Mycenae in the Peloponnese and the need to plan for their conservation, preservation and display. The most prominent example of the use of seismic risk assessments (which incorporate

---

[24] Seismic attenuation defined as 'the loss of seismic wave amplitude with distance' (Reiter 1990)

seismic hazard) can be seen in the conservation programme of monumental architecture such as at the Parthenon, Athens, (Korres *pers comm.*) and the temple of Apollo Epikourios, Bassai, (Cooper, 1996).

In archaeoseismological appraisals, however, it is the propensity to, and effects of, earthquakes (and associated phenomena) at a site that are of primary importance. For that reason it is the consideration of seismic hazard (in addition to the seismic deformation) both at a regional and a site specific scale, that allows for a realistic assessment of the most likely mechanism of any observed damage. For example, the presence of crushed skeletons (in a non-ritual burial situation) and bowed or offset walls, in an archaeological site which is located within an area of extremely low seismic hazard, *cannot be simply* equated to earthquakes on the grounds of the field observations alone. The fact that the site lies within an area of extremely low seismicity should alert the archaeologist that it is possible that both secondary seismic and, non-seismic, mechanisms of destruction may also be present.

Seismic hazard can be measured in terms of the expected intensity of strong ground motion at a given point, the deterministic evaluation of peak ground acceleration, and the probabilistic measure of reaching peak ground acceleration. Additionally, seismic hazard assessment can be carried out at a number of scales from the regional and small scale covering a large geographical area, to locality or site specific, large scale. In this research, the seismic hazard has been considered both regionally and, where possible, site specifically (i.e. the archaeological sites used within the study) in terms of (1) the expected intensity, which is essentially a qualitative measure or 'feeling' of the earthquake; (2) the calculated peak ground acceleration based on the 1894 earthquake parameters, i.e. M = 7.0 (Ambraseys & Jackson, 1990), and (3) the probabilistic evaluation of likelihood of reaching peak ground acceleration within two separate time intervals, taken from the work of Ganas (1997).

*Macroseismic Intensity - Regional, Small Scale*
Regional, or small scale, hazard measurements are based on the observation that the intensity of ground acceleration attenuates (decreases) with distance from the point source (hypocentre or focus) of the earthquake (Reiter, 1990). Papaioannou (1984) (in Papazachos & Papazachou, 1998) proposed the following relation between distance (Δ) in km (from the seismogenic source), earthquake magnitude (M) and the macroseismic intensity (I) for shallow focus earthquakes (i.e. with foci in the upper 8 km of the crust) in central mainland Greece as:

$$I = 6.95 + 1.18\,M - 4.5 \log (\,\Delta + 17\,)$$

An assumption of characteristic earthquake behaviour (i.e. rupture nucleation at the centre of the segment) and a 8km hypocentre (which is considered the outer limit of

shallow focus events in mainland Greece) places the epicentre of an earthquake event on the Atalanti fault with similar parameters as the 1894 event in the region of Skala. Figure A6.1 and A6.2 in Appendix 6 show the derivation of the distances Δ and R using the geometry of the fault.

*Probabilistic Method*
The probabilistic method of seismic hazard assessment not only determines the peak ground acceleration expected at specific points but allows for the uncertainties, such as not knowing exactly when an earthquake will occur, by assigning numerical probabilities. This is done by defining the geometric relation of the site to known seismogenic sources, calculating the recurrence interval (usually based on the palaeoseismic history of the faults), assigning the earthquake effects (i.e. maximum ground acceleration) and calculating the probability of exceedance of maximum ground acceleration. The values used in this investigation are from Ganas (1997) and have been derived using the earthquake parameters similar to those of the 1894 Atalanti earthquake (with a characteristic earthquake behaviour) using a semi-statistical computer software program, FRISK.[25] Although values are computed specifically for the sites of Kynos and Kyparissi, no calculations are given for Alai and computation of such is beyond the scope of this thesis. However, based on the similar properties of Alai to Kynos (i.e. R values, recurrence interval, bedrock material, etc.) it is reasonable to suggest, for the purposes of this investigation only, that Kynos and Alai have similar probabilistic values.

The hazard map values (inferred [in italics] for Alai) are given in Table 4.7. The negligible difference in values for each site in relation to the horizontal distance from the source, i.e. the Atalanti fault, is a result of the marginal difference in R values where Kynos lies closer to the hypocentre due to the inclined geometry of the fault. So, it can be said that with a 2000-year recurrence (from a palaeoseismic history that assumes that the 426 BC earthquake, as well as the 1894 Atalanti event, occurred on the Atalanti fault)[26] is 0.05g at Kynos and 1000 at Kyparissi with a 37% probability of exceedance (i.e. low probability). The data, at this scale, show that ground acceleration reduces with distance from the source (as expected for regional scale) and that there is negligible attenuation of seismic energy, based on source geometry.

In terms of an archaeoseismological appraisal it can be said that there is a very low probability of two or more seismic events occurring on the Atalanti fault within a time period of 1000 years. Therefore, if two destruction

---

[25] FRISK is a USGS computer program used to calculate seismic hazard.
[26] The assumption that the Atalanti Fault is also the seismogenic source of the 426 BC event is not unanimously accepted (Buck & Stewart *2000*), but is considered here due to the limitations of the available data on seismic hazard.

horizons are observed with the stated time period, it is reasonable to infer that one destruction is (1) the product of an earthquake occurring on an alternative seismogenic source, or (2) the product of an alternative destruction/damage mechanism.

In order to obtain a detailed archaeoseismological assessment of the effects of earthquakes on a site it is necessary to consider all the known seismogenic sources within the region. It should be noted that there are a number of other possible seismogenic sources in the vicinity, not least the Kammena Voula fault. However, as noted above, it may not be possible to differentiate which fault is the culprit for a destruction layer associated with ground acceleration. It is also important to note that these regional, coarse resolution, calculations do not allow for site effects (i.e. they are based on bedrock parameters), which will have a greater bearing on the effects of ground acceleration on structural remains (Reiter, 1990).

*Deterministic Method*
This method assumes 100% probability of a seismic event occurring, and calculates the peak ground acceleration (for bedrock) at a given point based on specified earthquake parameters, such as magnitude, R value, etc. In this investigation the calculations (Appendix 6) are made using earthquake parameters similar to those assumed for 1894 Atalanti event,[27] to provide the maximum ground acceleration values that could be expected to have affected the sites in both the 1894 and previous events.

*Local Site Effects*
A more detailed study of the local site effects provides a finer resolution in the hazard assessment for each site. Angelides (1992) carried out a geotechnical appraisal of the Atalanti plain and published his findings (in Greek) in a report accompanied by a series of geotechnical maps (see section 4.3). However, in discussing the properties of the sediments he uses the term 'seismic risk', and it is not clear from the text whether he uses the terms 'risk' and 'hazard' interchangeably as is often the case with specialists outside seismic hazard analysis. As archaeoseismic appraisal is concerned with the likelihood and scale of damage or vulnerability of archaeological remains, based on the site effects, it is reasonable to use the seismic risk delineation provided in the geotechnical data, based on the premise that more damage will occur in sediments that amplify ground motion. It seems likely, given the context of his work, that the terms 'risk' and 'hazard' have been interchanged, and 'risk' here is actually used in the context of hazard. The relevant data, i.e. sediment cohesion, ground water depth, and resulting seismic risk classification derived from the geotechnical map, are tabulated for ease of reference in Table 4.1, and seismic hazard evaluation is shown in the corresponding diagram at Figure 4.9.

The Malesina peninsula and Theologos bay areas were not included in Angelides' study, therefore the seismic hazard for these areas was extrapolated where possible by correlating bedrock lithologies. However, correlation was not possible for the surface deposits of the Theologos plain, as this would have required specific and detailed sediment analysis and calculations. Therefore the data for this area are, at present, missing and represent a lacuna in the dataset.

*Intensity and Ground Acceleration in the Atalanti Study Area*
In summary, the interplay of seismic hazard calculations at a variety of scales (small scale regional data and local, site specific date) allows a more reasoned assessment of the mechanism of any observed damage in archaeological remains. The results in Table 4.7 show that seismic hazard in the Atalanti plain (following the relatively recent rupture on the Atalanti fault in 1894) is relatively low. Additionally, the deterministic calculations of peak ground acceleration and intensity that can be expected for a large magnitude event on the Atalanti fault provide a reference dataset when assessing supposed seismic damage at specific sites, and thus by analogy the type of damage that may be observed (though, as noted above, care should be taken when correlating damage to modern structures with that observed in ancient remains).

At a regional inter-site level the seismic hazard of the Atalanti region is shown to be medium (a 37% probability of exceeding peak ground acceleration as indicated for each site in the results table above) for a 1000-year recurrence interval; (see the calculations at Appendix 6 (from Ganas, 1997). At a local intra-site level, it can be seen that most of the plain will experience intensity VIII to IX ground acceleration (calculations based on bedrock, with an epicentre in the Skala area), which is capable of producing moderate to major damage such as fractures of walls, collapse of weak structures, toppling of chimneys, etc. (see the Modified Mercelli scale of damage at Appendix 7). Additionally, it should be noted that this figure will be enhanced by the site effects, especially along the coastal regions (see Figure 4.9), where unconsolidated sediments and high ground water tables will amplify the seismic wave velocities.

The site at Kyparissi is particularly vulnerable to the effects of ground acceleration, being located on less than 10 degree slope within 5m of the supposed Atalanti 1894 ruptures, with foundation material having low cohesion properties. Similarly, the coastal margins of Almyra and Palaeomagasin are expected to experience amplified seismic shaking. This has been supported by field observations in the coastal area of Palaeomagasin where ashlar block structures (undated) show deformation which is probably the result of either liquefaction (a product of shaking of coastal sediments) or slumping, also a secondary effect of seismic excitation

---

[27] Magnitude etc. from Ambraseys & Jackson (1990).

| | Kynos | Kyparissi | Alai |
|---|---|---|---|
| Distance from fault (Δ) | 10km | 1km | 10km |
| Distance from focus (R) | 9.13km | 9.2km | 13.6km |
| Intensity (MM Scale) | VIII - IX | VIII – IX | VIII - IX |
| Peak Ground Acceleration (Theodulis) | 827.4 gals | 827.4 gals | 630.9 gals |
| Peak Ground Acceleration (Makropoulos) | 658 gals | 672 gals | 523 gals |
| Seismic Hazard (Ganas 1997) | LOW | LOW | LOW |
| Site Effects (Angelides 1992) | LOW | HIGH | *LOW* |

**Table 4.7**: Summary of the seismic hazard and ground acceleration calculations for the three archaeological sites in the Atalanti Plain.

(see chapter one for examples of the type of observations that have in the past been used to infer seismic mechanism). Both Kynos and Alai are founded on bedrock, and therefore are not affected by local site effects.

## 4.7 Human Factors

As noted in the introduction the purpose of this empirical database is to collate information that will be useful to archaeoseismological appraisals. As noted in chapter 2, an understanding of both the natural and human environment are central to the construction of a rational argument for or against the designation of a seismic mechanism for observed damage to archaeological remains. It is suggested that the human factors should also be considered at several different resolutions, from the broad scale (coarse resolution) national and regional understanding of the main events to the specific historiographies of the sites under review (high resolution). For the purpose of this research, two archaeological datasets have been collated from a number of sources.

The first is a general archaeological history of mainland Greece with key events, both in time periods and date specific, and is located at Appendix 8. The second are résumés of occupation and excavation histories of the sites these will be discussed in chapter 6 and have been incorporated within this chapter at section 4.7.2.

### 4.7.1 General Chronology

The information provided in this section is derived from an amalgamation of material from several source books on the history of Greece (Fossey, 1990; Demand, 1996; Martin, 1996; Morkot, 1996;). Due to the presence of a variation of dates between the sites for named occupation periods, i.e. Palaeolithic, Neolithic etc., generalised date ranges have been used to prevent over complication of the dataset by providing individual site-specific chronologies. At the end of the general discussion of the information, a small regional section for Opuntian Lokris has been included for areas where there are available data. For the later periods which overlap with the availability of textual information, the dates used tend to correlate across the source books used and these dates have been maintained in this section: for example the death of Cleopatra in AD 30.

### 4.7.2 Site-Specific Occupation Histories

The large number of archaeological sites that lie within the field area, the requirement of an official permit to use information from sites that are not fully published,[28] the paucity of published literature, and the time constraints imposed by the research project dictated that only a sample of sites could be considered in depth. With these feasibility considerations in mind, and considering the initial questions of the research strategy, the following

---

[28] Permit granted by the Ministry of Culture with the support of the regional Ephorea, Dr Ph Dakoronia, excavation directors, and the British School at Athens.

three sites were selected for more detailed examination Kynos, Kyparissi (Opus), and Alai. The sites were chosen for the following reasons:

- their geometrical location in terms of distance from the main fault trace and their position along strike of the segment
- collectively long occupation histories, but with different periods of excavation focus

To allow for comparative analysis, the information for all three sites is presented in the following format: a brief introduction to the locality, followed by a chronological listing of references to the sites with a brief discussion of any archaeological excavation, and finally a brief outline of the known occupation periods derived from the available references.

*Kynos*
Ancient remains have been found, in two localities in close proximity to one another, approximately 2km north of the modern town of Livanates. A hill site founded on a small flat topped hillock (with the modern appellation of Pyrgos), and a slight, and certainly less obvious, coastal installation immediately below the hillock, constitute what is considered to be the Homeric settlement of Kynos, harbour town to the Opuntian capital of Opus. The apparent correlation between the modern geographical position of Kynos (at the north of the Gulf of Atalanti and the Opuntian plain) and that described by the ancient texts (specifically Strabo, who provides quantitative distance measurements between several sites) resulted in the unquestioning identification of this site by the early travellers (Gell, 1819. p 229; Dodwell, 1819, p58; Leak, 1835. p 182).

Although the visible remains have ensured that the site has not been forgotten over the centuries, the lack of published material would suggest that the site has not been considered archaeologically significant in the past. Short commentaries on brief reconnaissance visits note the high density of pottery sherds, ranging from Mycenaean to Hellenistic (see Hope Simpson and Lazenby, 1970; and Fossey, 1992; for references). Towards the end of the 1970s, an excavation campaign directed by Dr Ph. Dakoronia at the behest of the Greek Archaeological Service was started on the hill site, and this continues today. Excavation findings have been presented in several publications including a paper discussing the Middle Hellenistic graves (Dakoronia, 1978, in German), the yearly reports in *Archaiologikon Deltion*, and a conference paper specifically discussing the possible seismic damage in the Mycenaean period.

An additional more general article that discusses the identification of the Pyrgos site with Homeric Kynos notes that 'the archaeological data do not support the ancient descriptions of Kynos as a big, important town with a well-developed terracotta industry' (Dakoronia 1993. p 125). This, when combined with the dis-

similarities between Strabos' distances with and modern measurements (even considering any coastline changes) between settlement sites, and apparent physiographic issues, should surely raise questions about the correlation of this site with Kynos.

Based on the present published material it is impossible to give a detailed site chronology, which awaits the completion and publication of the archaeological investigations. However, both the surface surveys and excavation finds to date indicate occupation as early as the Middle Helladic through to the late Byzantine. However it is important to stress that Dakoronia specifically notes that not all periods are represented in terms of habitation, especially on the hill itself, which has a surface area of no more than 17 acres (1993, p125).

*Kyparissi/Opus*
The archaeological site considered to be the locality of Opus, ancient capital city of Opuntian Lokris, is located 2km south-east of the modern (post-1894) hamlet of Kyparissi, adjacent to a large brewery. Again this is a dual locality site, both localities lying within 200m of the proposed trace of the Atalanti fault along the Quaternary/bedrock contact. Well-preserved polygonal walls trace the outline of an acropolis atop the ridge-like, 300m high Kokkinovrakhos hill, whilst an associated settlement site is thought to be located on the lower northern and north-western flanks of the same hill: the latter's position is indicated by the high density of potsherds on the surface and the presence of a number of subsurface structural remains, especially in the vicinity of the chapel of Agios Ioannis. The search for this ancient city, noted by Strabo in the *Geographies*, occupied the minds of many of the early travellers and has certainly been hampered by the conspicuous lack of any surface structural features such as walls.

The Kyparissi locality for Opus is favoured by most early travellers, including Gell (1819. p 229), Dodwell (1819. p 58), and Leake (1835), and by later scholars, for example Lolling (1879), and is supported by the wealth of ceramic sherds ranging from the Bronze Age through to the Archaic period that have been found in that area (Hope Simpson & Lazenby, 1970, Dakoronia, 1993). Although the acropolis site had been surveyed and the architectural remains described several times since the 1880s (see Fossey, 1990 for references), it was not until the spring of 1911 that the first archaeological investigation of the supposed settlement area was made under the patronage of Prof. C. D. Buck of the University of Chicago, USA. A single publication by the excavation director, C. Blegen (1926), outlines the finds from the enterprise and specifically notes the discovery of a stylobate[29] attributed to the 5th century B C.

Further excavations carried out in 1978 and 1979 by the Greek Archaeological Service, under the directorship of

---

[29] A continuous horizontal course of masonry that supports a colonnade

Dr Ph. Dakoronia, pushed back evidence of occupation to the Archaic period (6th century B C). However, the contextual dating is tenuous and relies heavily on architectural terracottas which require re-investigation in light of the recent re-examination and excavation of the site at Alai, which throws doubt on the dating of the architectural terractottas (see Buck & Stewart (2000) for further discussion on this point). Other than the extensive literature pertaining to the apparent seismic damage, there is little published material for this site; thus it is not possible to list further details on the occupation history of what appears to be a potentially large settlement site.

*Alai*

The acropolis site of Alai (ancient Halai) is located on the south-eastern shore of the Bay of Atalanti, in the vicinity of the modern fishing village and tourist destination of Theologos, approximately 6km NE of the Atalanti fault. There are above ground relics and text references by several classical authors, the most prominent of which describes the sacking of the town at the command of the Roman general Sulla in 85 BC (*Plutarch ix. 24. 5*). The final ancient reference to the site is found in Pausanias (Walker and Goldman 1915) where Alai (Halai) is described as long forgotten and desolate. The next reference is until the identification and appellation by Leake in his travel memoir *Travels in Northern Greece. Vol. 2*, published in 1835. As noted previously in section 4.4.3, Lokris was visited infrequently by early travellers, resulting in few 19th and 20th century descriptions or studies: Alai is no exception.

The first archaeological excavations were carried out between 1911 and 1914 under the auspices of the American School of Classical Studies in Athens and the directorship of Hetty Goldman and Alice Walker. The wealth of finds at the acropolis site and adjacent necropolis (consisting of over 280 graves) resulted in a further four excavation seasons between 1921 and 1935 with the assistance of Virginia Grace (Coleman, 1992). Publication of these early campaigns is limited to a preliminary report of the first four sessions (Walker and Goldman, 1940), the completion of two doctoral dissertations on the finds of the tombs by both Goldman and Walker, and a few finds from the acropolis site by Goldman (Coleman, 1992).

More recently the remains have been the focus of the Cornell Halae and East Lokris Project (CHELP), an American interdisciplinary archaeological and environmental project under the directorship of Professor John Coleman[30]. The interdisciplinary ethos of the project is highlighted by the number of non-archaeologists (e.g. geologists, computer scientists, anthropologists) that have contributed to the project and the progression of activities

from surface survey (1988-1989) to excavation with contemporaneous consolidation work (1990-present). Contrary to the majority of modern archaeological investigations, several preliminary reports have been, or are currently being, published, notwithstanding the prodigious volume of finds and associated research (classification, analysis). Additionally, the research is fully accessible on a regularly updated website.[31]

To date there have been no finds at Alai to suggest Palaeolithic occupation and the earliest levels founded on 'sterile ground'[32] date from the Early Neolithic (c.5900-5800 BP). Evidence of uninterrupted habitation continues through the Middle Neolithic (c.5700 BP) but did not extend beyond the early Late Neolithic (5400-5300 BP). Following an occupation hiatus of approximately 4,000 years, Alai was re-founded in the Archaic period (c.700-500 BC). Occupation continued almost uninterrupted through to the present, though there is little material from the Classical period (5th & 4th centuries B.C) which may indicate a population in decline. The Hellenistic (approx. 3rd to 1st century BC) is well represented, with many of the visible structures dating from this period, indicating an extensive occupation which culminated with the sacking of the town by Sulla in 85 BC. Reoccupation in the Late Roman to Early Byzantine period is indicated by 'fill' deposits and the presence of Byzantine architecture in the form of walls and a Basilica style church. Continued habitation into the Late Byzantine and Early Frankish periods is, evident from the presence of chapel alterations, graves and dated coins.

## 4.8 Conclusion

There are many problems in establishing an empirical database such as the one in this chapter, for example 'gaps' caused by incomplete and/or unavailable data sets, or the difficulties of using data and information that has been published in a variety of languages. However, these are minor details which true interdisciplinary collaboration can to a certain extent, mitigate against. Question: without turning back can you recall the seismic hazard of the Xerovouni plateau, or the archaeological site of Kyparissi? Probably not, and the reason is that since reading the geotechnical section other datasets, from several other disciplines, have been added. It may be that a pause in reading was taken to complete another task, and although there is a vague memory of where the sites are geographically, it is not easy to re-orientate them in relation to the information given by the geotechnical section. Thus, it can be seen that the main problem with employing numerous datasets in interdisciplinary research is the volume and variety of data required to tackle research questions. Not only does it need to be read, and understood, but the information needs to be

---

[30] Permission to use information relating to the CHELP project specifically unpublished (at the time of writing) radiocarbon data and information contained in the WWW site was given by the director, Professor John Coleman.

[31] http://www.halai.fac.cornell.edu/chelp/home.htm.

[32] Ground that shows no previous evidence of occupation; in the case of Alai, the Earliest Neolithic is founded on the Neogene deposits.

retained in a manner which allows for ease of recollection at a later date. It is this problem that is hinted at by Nikonov (1988) in the concluding discussion of his proposed methodology for archaeoseismology, and which will be discussed further in the subsequent chapter.

# Chapter 5: Geographical Information System[33]

The main problem with archaeoseismological research which has been noted in the work of previous authors (for example Karcz & Kafri, 1978 and Nikonov, 1988), and highlighted in the conclusions to Chapter 4, is the number and diversity of datasets required to carry out a true interdisciplinary appraisal. Inherent in this problem is the added difficulty of retaining and recalling the necessary information when evaluating each observation or occurrence of supposed seismic damage. In the light of these problems a Geographical Information System (GIS) was employed to provide a means to store and simultaneously recall multiple data sets. With an increased understanding of GIS it became apparent that the main value of the system lay in the ability to manipulate data to produce information which could be used in a decision making process. In this context, the focus of the application of the GIS switched from primarily a recall and output facility of inventoried data layers to a tool to facilitate the qualitative assessment of archaeological remains as indicators of neotectonic activity through the manipulation of the digitised data layers contained within the empirical database.

## 5.1 Geographical Information Systems

An 'information system' is essentially a series of related or connected operations that moves the operator (be it research director, planner or project co-ordinator) from the initialisation stage of a project to data collection, data storage, manipulation and analysis, to output, and ultimately to the application of the results in a decision making process. This may, in turn, return the user to the planning stage (Star & Estes, 1990). Chrisman (1997) represents this inter-related and wholly dependent procedure as a series of 'nested rings' where each subsequent ring (or operational level) builds upon and supports the preceding action and subsequent results. If one ring is removed, the hiatus in the system punctuates the procedure, preventing a full realisation of the system potential.

Clearly, information systems are not simply a product of the development of computers and are not necessarily technology dependent, although the use of powerful hardware has certainly raised the profile of GIS systems due to the automation of calculations which has vastly improved the speed of producing results.

The term 'geographical' indicates the type of data to be considered within the information system, i.e. spatial.

However, these do not necessarily have to pertain to *real earth* parameters; a system which deals with arbitrary co-ordinates in an hypothetical town or terrain should not be excluded because it lacks real earth geo-referencing.[34] Therefore, broadly speaking a GIS is a tool that allows for the processing of spatial data into information (deMers, 1997), where data are defined as facts, or numbers representing facts or measures, and information is the meaning derived from multiple facts or numbers (or single pieces of data) (Davis, 1996).

In the past decade, there has been a huge up-surge in the use of GIS both in academic research and in industry. The breadth of applications in geology and environmental science are too numerous to list here, but often they have an economic common denominator such as disaster management and/or risk assessments; for example, they may be used in projects concerning deforestation, desertification, or floods. The introduction of GISs in archaeological research has been noted as being somewhat slower (Allen, 1990). Kvamme (1990) suggests that the more likely reason is archaeologists' "intense focus on the past which perhaps inhibits us [the archaeologists] from looking forward" in (Gaffney & Stançiç, 1991). However, Gaffney (1991) points out that the lack of uptake is more probably to do with (1) the cost (including the initial outlay) of the hardware and software required to both set up and maintain a fully integrated GIS, and (2) the lack of computer literate archaeologists to make maximum use of such a system. Additionally, a third factor can be included: the cost (in terms of time and money) of obtaining the required digital format data necessary to establish a GIS based project.

To date the main focuses of GIS based archaeological projects fall into two categories: (1) academic research programmes which appear to focus on spatial distribution or survey data (with or without physical environmental factors), and, (2) cultural resource management (CRM) projects which are often under the remit of local authorities or private heritage companies or groups. Examples of both groups can be found in the dedicated volumes *Interpreting Space: GIS and Archaeology* (Allen *et al.*, 1990) and more recent *Archaeology and Geographical Information Systems* (Lock & Stançiç, 1995).

### 5.1.1 Fundamentals of GIS

To the non-specialist, perhaps the most prominent feature of a GIS is the relatively easy production of high quality visual representations of information in a variety of formats, including maps, graphs and diagrams (for example, Figure 5.14 [orthographic projection of the

---

[33] The reader should be aware that this chapter was originally written in 1998 and has not been updated to include any of the subsequent new material especially in archaeology and GIS. It is included here simply as background to the work carried out for this research in Chapter 6 case studies.

[34] Geo-referencing refers to the process of locating spatial information with respect to a common frame of reference, which in GIS is usually a specified co-ordinate system (Burrough & McDonnell, 1998).

Almyra area]). However, this should not lead to the misconception that a GIS is simply a cartographic process analogous to computer assisted cartography (CAC) or computer assisted design (CAD). The *fundamental difference* between GIS software and automated mapping or drawing packages is that a GIS *stores* raw data (as co-ordinates plus attributes) *within the system*, and uses combinations of this stored data to *create information* from which integrated decisions can be made.

This major difference can be seen as the result of a paradigmatic shift in mapping concepts (deMers, 1997) from simple visual display of several data layers (e.g. contained within a traditional sheet map format) to analytical communication and the desire of the operator/user to be able (at any point following input) to access, manipulate, analyse and output data (and combinations of data) on a *user defined* basis. Essentially this changes the perception of cartography (plus associated disciplines) and the products of cartography (i.e. the visual portrayals of data) from static data display (sheet maps) to a dynamic information provider (computer display with optional printing).

### 5.1.2 GIS in Archaeology

As noted previously, archaeology is primarily concerned with people and material culture. An inherent problem with archaeological thinking and practice (i.e. the interpretation of behaviour and material culture over time) is the requirement of a systematic and concurrent consideration of three-dimensional data, i.e. space, time and form (Allen *et al.*, 1990). The fundamental properties of GISs, (spatially referenced databases) were originally viewed as a possible solution to this problem. Therefore, initially GIS was seen as a tool to allow archaeologists to reference observations spatially with attributes (such as artefact and provenance or, similarly, site with the environment) in a way that would subsequently facilitate the comparative analysis and/or mathematical manipulation of distributions in a spatial context (i.e. the environment). Additionally, the ease and flexibility of revision, analysis and, perhaps more significantly, product (or information) output were all contributing factors in the adoption of this non-indigenous tool.

Harris and Lock (1995) note that the use of GIS in archaeology is at present going through a growth trajectory characterised by stages of increasingly sophisticated of application. In the first stage, GIS was used primarily as an inventory tool, allowing a spatial element to be incorporated into data cataloguing. The most common application of this can be seen in localised archaeological records held under the statutory and regulatory responsibilities, for example the Sites and Monuments Records required by law in Britain to be kept by the local authorities, which are used extensively in planning procedure. Within this context the primary function of the GIS is as a cultural resource management

(CRM) tool. However, even within this basic objective the achievement of the theoretical ideal of a fully integrated, spatially referenced database is a tedious, time consuming and costly activity that raises a number of issues relating to the difficulties and deficiencies that accompany multi-dimensional data: namely, the theoretical problems associated with switching away from traditional archaeological interpretations based on two-dimensional abstraction of three dimensional observations, for example object and location with height, time or depth (Harris & Lock, 1995).

The second stage focuses on the use of GIS in analysis, and is typified by applications within landscape archaeology, and/or research that are primarily concerned with human land use and the connection of cultural remains (and corresponding theory) with natural factors, i.e. environmental archaeology. In the early stages this has mainly taken the form of predictive modelling of site catchment in relation to site location, which often includes simple analytical queries such as line of sight models for inter-visibility between settlement sites, for example the Hvar Project (Gaffney & Stançiç, 1991). A straightforward explanation for this localised expansion of GIS within landscape archaeology is that cultural landscape data and the geo-referenced environmental data necessary for this type of research lend themselves well to integration and manipulation within a GIS, due to the presence of a spatial common denominator. Additionally, the output medium (i.e. cartographic representations) equally lend themselves well to analysis and interpretation within landscape archaeology. Latterly, and with the rise of cultural environmental archaeology, this stage has been expanded and GIS has become the pivotal tool to allow the integration of diverse datasets ranging from topographic and environmental data to site distribution, location and evolution. This has resulted in the attainment of archaeological site interpretations within their settings (i.e. the environment) which was initially advocated by Vita-Finzi (1978) in his book *Archaeological Sites in their Setting* and is well demonstrated in the Hvar Project (Gaffney & Stançiç, 1991).

The third stage, and one which is not well defined (Harris & Lock, 1995) is the use of GIS in an integrated decision making process. The York Archaeological Assessment (YAA) is an excellent example of how a GIS has moved from the initial inventory and basic analysis stages to being used in the management of the archaeological heritage under the city of York. One of the key goals of the YAA is to 'allow detailed querying of archaeological interventions within the city in relation to modern land use and cityscapes as well as the mapping and prediction of subsurface archaeology' (Miller, 1995). What places this particular project in the final stage is that a number of different data layers (including archaeological, topography, land use, and statutory protected areas) are used to create multi-dimensional representations of the area, including deposits and depositional processes within the physiography (Miller, 1995), which are then used in the management of the cultural resources.

Throughout this increasingly sophisticated use of GIS in archaeology, it has become apparent that there are a number of possible pitfalls (mainly methodological and theoretical) for both the unwary (and wary) which are described by Gaffney *et al.* as "potentially restrictive to the development of archaeological thought" (1990. p 211), and which should be considered throughout GIS based projects. These are:

- that the propensity to derive aesthetically pleasing output products may inform the direction of investigation or be used to justify [inappropriate] research, which may result in
- the repetitive confirmation of obvious relationships (derived pre-and post-GIS) in other words 'stamp collecting', and that the previous two points may result in
- an exposition of environmental determinism or functionalist theory[35] within archaeological interpretations.

(Gaffney, 1990. p 211)

Clearly the first two points are not exclusive to archaeology, and are equally applicable to any research that employs a GIS, which by definition is essentially inter-disciplinary in nature.

Although the use of GIS in this research originates in the first stage, the application of modelled seismic data falls loosely into the final stage of integrated decision making, in that it (1) uses a diversity of spatial and temporal data, (2) derives information from assimilated datasets and, more specifically, (3) facilitates a qualitative investigation of traditionally subjective material by creating a 'neutral' model from which to make a comparative analysis of subjective observations, thus testing the working hypothesis.

## 5.2 GIS Methodology for Testing the Working Hypothesis

Previously, archaeoseismological appraisals have focused on attributing earthquake damage through identification, either using a criterion or check-list approach, or from personal experience (see Chapter 2). As noted in the methodological discussion in Chapter 3, this renders each criterion (or check-list) and each specific identification of seismic damage highly subjective, and therefore unsuitable for making value judgements through simple cross-site comparisons. It is suggested in this work that to obtain a meaningful comparative assessment of

archaeological field observations as proxy indicators it is necessary to include a known standard or neutral 'reference' point.

As it is the archaeological data which are under scrutiny as proxy indicators, it is the archaeological field observations that need to be assessed in the light of what could be expected geologically from earthquakes of known parameters. This can be achieved by comparing the observations with expected geological phenomena derived from constructing an impartial, albeit theoretical, model. Thus, if specific archaeological evidence is in agreement with the theoretical model, it is reasonable to infer that a specific archaeological observation could, potentially, be a useful indicator.

It is proposed here that the use of theoretically derived models could provide the working hypothesis which essentially fulfils the third element of the strategy proposed in chapter 3. A qualitative assessment of archaeological (proxy) seismic data can then be made by a comparative analysis of the observed and expected phenomena. Due to the disperate nature of the data involved in the creation of the expected seismic phenomena models, the use of GIS in terms of storage and manipulation of the data layers, would enhance their production. Additionally, the creation and use of neutral models prior to archaeoseismological fieldwork is proposed as a means to reduce the subjective identification of future appraisals. Further, if archaeological observations are found to be useful indicators, the derivation and inclusion of these reference models to the interpretation of observations may in the future assist in obtaining more quantitative earthquake information from seismically damaged archaeological remains, thus providing a more reliable data source for seismological research.

### 5.2.1 Theoretical Models of Seismic Effects in Lokris

To create a model of the effects of an earthquake in a region, or at a particular site, it is necessary to 'deconstruct' the seismic event and to model (separately) the two principal effects (i.e. the deformation and ground acceleration) of the earthquake. The effects of the earthquake can then be modelled by using a GIS to bring together the separate elements in one output image.

*Deformation Ellipses and Zones of Influence*
The amount of subsidence or uplift in an archaeological site can only be quantified when there is a laterally continuous datum, for example mean sea level as indicated by the coastline (see Figure 5.1). Further, even when a datum exists, the seismic element of deformation may be difficult to identify, or conversely it may be inappropriately overemphasised depending on the archaeological remains used and the experience of the archaeologist.

---

[35] Theoretical environmental determinism in archaeology regards past and present cultures as functions of, or shaped by, environmental pressures such as terrain, climate and natural phenomena such as floods, earthquakes, etc. (Gaffney & van Leusen, 1995. p 367).

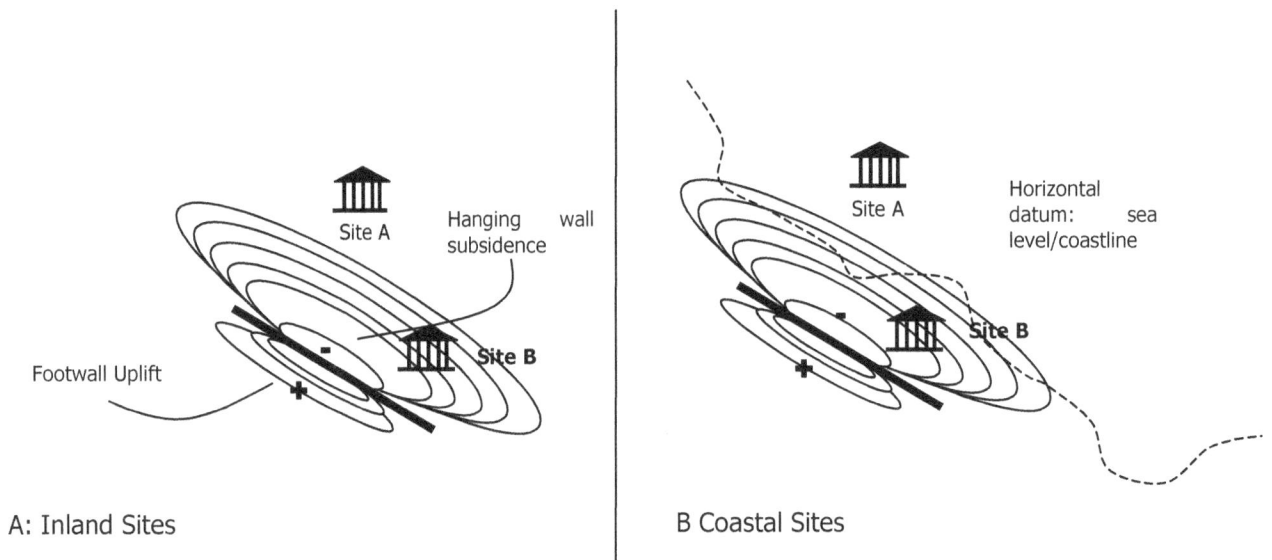

A: Inland Sites

B Coastal Sites

**FIGURE 5.1:** A SCHEMATIC REPRESENTATION OF HOW DEFORMATION ELLIPSES CAN BE USED TO ASCERTAIN WHETHER SUBMERGED/UPLIFTED REMAINS COULD FEASIBLY BE THE RESULT OF NEOTECTONIC ACTIVITY. THE LACK OF A LATERALLY EXTENSIVE VERTICAL DATUM IN INLAND SITES (A) MAKES IT DIFFICULT TO IDENTIFY ANY VERTICAL MOVEMENT BETWEEN SITES A & B. THIS IS ADDRESSED IN COASTAL SITES (B) BY THE PRESENCE OF THE MEAN SEA LEVEL MARK AT COASTAL ARCHAEOLOGICAL SITES. (NOT TO SCALE.)

Therefore, in terms of archaeoseismology, it is the spatial extent of an area of deformation associated with an earthquake (or sequence of earthquakes) that is of primary importance as a guide to interpreting whether relative vertical movement in a site is the result of co-seismic slip or the product of other alternative mechanisms, for example coastal slumping or a eustatic change in sea level. Within this context, the original mathematically derived quantitative deformation ellipses associated with co-seismic slip on a fault can be regarded as discrete geographical areas that have been affected by deformation (either positively or negatively), or a *Zone of Influence (ZOI)*.

Once derived, *ZoI's* can be used to assess the validity of the commonly used practice, especially in coastal research, of using archaeological sites as indicators of neotectonic activity (in the form of permanent vertical displacement of the physical landscape). This can be achieved simply, by comparing the geographic location of the archaeological site in relation to the spatial extent of the zone of deformation (subsidence and uplift) for nearby faults, or the ZoI. For example, if archaeological remains are observed below sea level and the mechanism of their submergence is cited as due to an earthquake, then this observation can be validated by noting whether the site falls within the derived ZOI calculated from specified earthquake and fault parameters.

At the point of writing primary data from deformation modelling were not available to this research and as such

it was not possible to directly import primary data into the GIS to use with the Digital Elevation Model (DEM) produced for this purpose (see Appendix 5 Figure A5.11) Therefore, it was necessary to derive ZoIs from adapted models by Ganas and Buck (1998) based on the original work of Ma and Kusznir (1995) and using standard modules within the GIS IDRISI to produce approximations that could be used to demonstrate the theory discussed above[36]

The original displacement ellipses were modified as noted below to reflect the parameters of the Atalanti (Lokris) fault:

- a large magnitude characteristic earthquake, on a
- rectangular fault 34 km in length with a 55° pure dip slip plane buried in an elastic half space[37]

As noted above, the GIS derived ZoIs are geometrical approximations and as such they are derived using a series of structural assumption both in the mathematical modelling and in the GIS procedure. These are that the seismogenic sources are composed of straight segments, i.e. there is no account of along-strike irregularities and that the seismic event is the maximum possible along each seismogenic source.

---

[36] The GIS method of construction can be found in section A5.3, in Appendix 5.
[37] An elastic half space is a two dimensional representation of the earths' crust

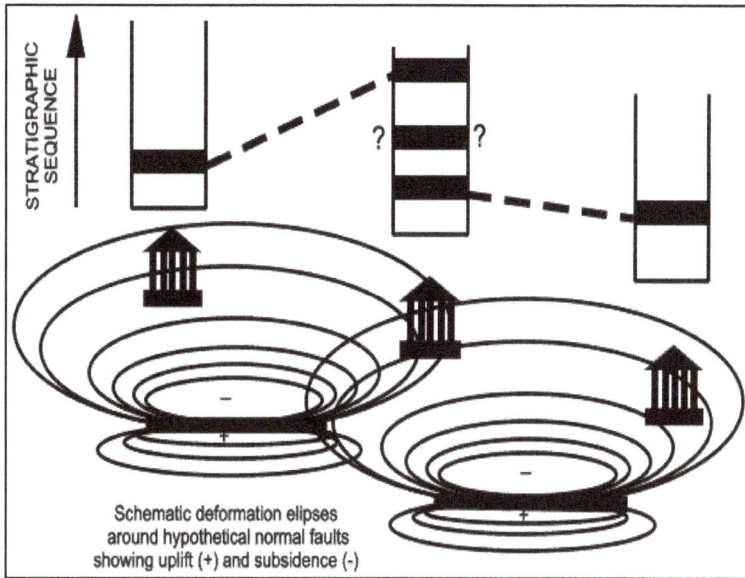

FIGURE 5.2: SCHEMATIC REPRESENTATION OF THE CONCEPT OF USING CORRELATION OF DEFORMATION HORIZONS TO CORROBORATE THE MECHANISM OF DEFORMATION AND CONSTRUCT A REGIONAL CATALOGUE OF SEISMIC EVENTS.

The assumption of the straight fault segment is valid due to the scale of the modelling. At 1:250 000 (or regional scale) minor deviations from linearity are negligible and can therefore be discounted. Additionally, the seven seismogenic sources that are considered as discrete fault plains with a general NW – SE strike showing no segmentation. The second assumption is reasonable because: (1) for archaeological appraisal it is necessary to know the effects of the maximum earthquakes, and (2) the 1894 Atalanti earthquake may have been a maximum magnitude earthquake (Skouphos, 1895). Allowing for these assumptions, the GIS enables the creation of:

- symmetrical distance pattern, showing no differentiation between areas of uplift and areas of subsidence respectively, with
- no tapering of deformation at segment boundaries.

It should be noted that the use of alternative earthquake behaviour models (for example, modified overlap (see section 4.2.5)) or fault architecture (i.e. non segmented or non linear fault planes) would produce different configurations of the ZoIs within the GIS. However, this would not detract from the underlying theoretical basis that the Zones of Influence associated with individual fault segments can assist in the interpretation of the likely cause of submerged/emerged archaeological remains. Further, direct importation of primary modelling results into the GIS would only expand the utility of the ZoI application, for example the extraction of pre and post seismic topography using rasterised graded ellipse and a DEM (see section A5.4.3 in Appendix 5).

### Correlation of Deformation Events
An enhanced interpretation of the mechanism of deformation can be obtained through the correlation of deformation horizons in stratigraphies of adjacent archaeological sites which are then compared to a theoretical model of seismic deformation in a region. This can be done by OVERLAYing the ZoIs for all seismogenic sources in a region on a digital image file (Figure 5.3) and then correlating the deformation horizons (i.e. relative changes in sea level) in sites that occur in a single ZoI and those that are situated in the overlapping sections of one or more ZoIs. This is represented diagrammatically in Figure 5.2. From this a deformation stratigraphy for the region can be constructed and where possible dated archaeologically.

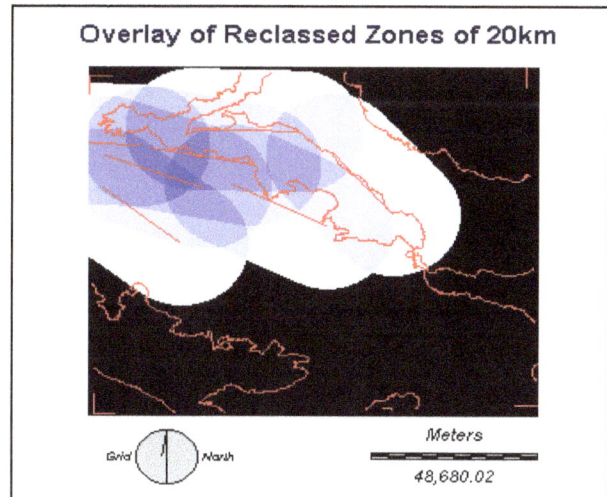

Overlay of Reclassed Zones of 20km

FIGURE 5.3: THE FIGURE SHOWS COMBINED ZOI (7 SOURCES). OVERLAPPING ZOIS INDICATE AREAS PRONE TO DEFORMATION FROM TWO OR MORE SOURCES. ACCORDING TO THIS FIGURE THE AREA WITH THE HIGHEST DEFORMATION SHOULD BE SOUTH OF KAMMENA VOURLA.

### Ground Acceleration
Consideration of the ground motion, in terms both of peak ground acceleration (both probabilistic and deterministic) and intensity, in archaeoseismological appraisals, allows a more complete understanding of the response, or likely response, of the ground (and indirectly the relics) to an earthquake of specified parameters. Analogy with the type of damage to structures in recent earthquakes will give an indication of the range of damage features that *could* be expected from an earthquake of a specific magnitude. However, caution should be exercised in making direct comparisons of modern damage (as described in the Modified Mercelli Scale at Appendix 7) field observations at archaeological sites, as there are numerous other factors that could

contribute to the damage observed: for example, construction materials, natural frequency of the buildings, foundation materials, etc. (Karcz & Kafri, 1978). Caution should also be exercised when trying to extrapolate the intensity or magnitude of an earthquake from archaeological field observations, as it is impossible to know whether the observed damage is the result of a single earthquake or of cumulative damage from several events (not necessarily from the same seismogenic source). Notwithstanding these limitations, the inclusion of intensity values in the model is justified as a basis from which to dismiss the improbable. For example, if maximum intensity values of III or IV are derived for a site that contains major structural collapse, it is reasonable to infer that the destruction cannot be attributed solely to the earthquake, and that other mechanisms are also in operation and should be considered, for example ground instability.

Similarly, probabilistic calculations of ground acceleration can only provide a basis from which to assess the likelihood that damages and destruction layers are attributable to seismic events (see Figure. 5.4). For example, if two destruction layers requiring high ground acceleration values are identified in an archaeological site, where there is a low probability of ground acceleration exceeding the required values within the time interval, then it is reasonable to infer either that one of the two destructions must have an alternative mechanism, or that they are the products of earthquakes on different seismogenic sources in the vicinity.

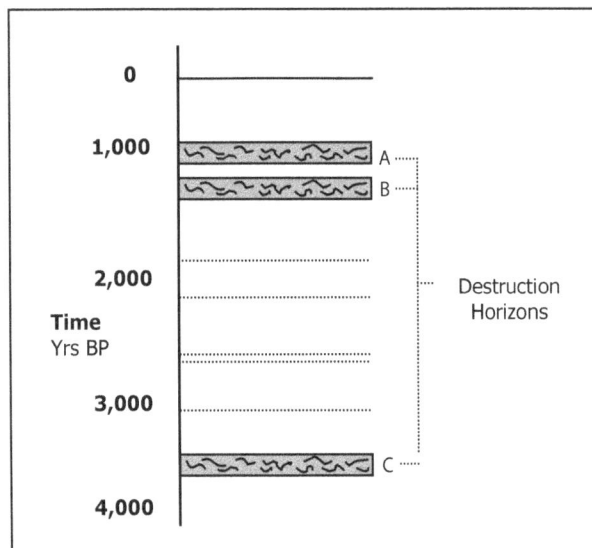

FIGURE 5.4: SCHEMATIC REPRESENTATION OF THE USE OF PROBABILISTIC CALCULATIONS OF SEISMIC HAZARD IN ARCHAEOSEISMOLOGICAL APPRAISAL. IT IS UNLIKELY THAT HORIZONS A AND B ARE THE RESULT OF EARTHQUAKES ON THE SAME SEISMOGENIC SOURCE, DUE TO THE VERY SHORT TIME INTERVAL BETWEEN THEM. THEREFORE AN ALTERNATIVE MECHANISM OF DESTRUCTION SHOULD BE CONSIDERED IN THE INTERPRETATION. HOWEVER, GIVEN THE LONGER TIME INTERVAL BETWEEN A (OR B) AND C, THEN IT IS REASONABLE TO INVOKE A SEISMIC MECHANISM (ON THE SAME SOURCE) FOR THE DESTRUCTION. (NOT TO SCALE.)

The ground acceleration values used in the construction of the effects model are noted in section 4.6.2 and are the product of derived information (Ganas 1997) and of primary data, from calculations by the author (see Appendix 6) using empirical formulas noted in (Papazachos & Papazachou, 1997).

*Site Effects*

As noted in section 4.6.2 (ground motion), local near surface conditions have a direct influence upon the velocities of seismic waves, and therefore should also be included in the construction of the models of effects. At the point of writing up this research the geotechnical maps from Angelides (1992) had not been digitised, therefore it has not been possible to include this dataset in the model and consideration of the surface conditions in the Atalanti Plain has been made through the use of sheet maps from Angelides (1990).

## 5.3 Discussion

In this chapter GIS has been discussed as a potential tool for use in archaeoseismological research in the form of 1) data storage medium, and 2) in the context of creating neutral seismic phenomena models as a means of testing a working hypothesis (as discussed in chapter 3). Due to a number of problems associated with data acquisition, it has not been possible, at the point of writing, to realise a fully operational GIS of the field area. Therefore, much of this chapter has to presented as a theoretical discussion of the potential of GIS as a tool for future archaeoseismological research rather than a demonstration of the capacity of an operational GIS.

### 5.3.1 Problems with the GIS.

At the outset of the research the GIS was envisaged as a tool (for archaeologists and geologists alike) to store and recall both numerical and text based spatial data. Using the existing licensed GIS package within the institution was justified at the time, as it appeared to fulfil the basic pre-requisites that were considered in the software selection. These were that as the GIS was being designed as a tool to enhance archaeoseismological appraisals it should be amenable (and to an extent self-explanatory) to both the initiated and uninitiated who had been provided with a short introductory course. So, for instance, the software should be Windows driven rather than command lines and that standard operations were available as modules rather than requiring specific user programming or querying. Outlay cost was also a consideration, not only for this project, but that the provision of a relatively low cost system is more in keeping with archaeological fieldwork budgets, or non-sponsored seismological research. To that end IDRISI version 1 for windows, in association with the older DOS version of IDRISI was used in this project. However, as research progressed it became clear that there were some fundamental problems with GIS.

Theoretically, the use of a GIS within archaeoseismology will enable the operator to move freely between the data sets, manipulating them to derive information which will assist in the appraisal (both locally and regionally) of supposed damage at a site. This ideal has not been fully realised in this thesis due to a number of drawbacks, both external and internal to the GIS selected, in terms of hardware, software and data.

The following subsections present a more detailed and specific discussion of the problems encountered with the use of the GIS. These can be divided into two groups, hardware and software problems. In the former, the basic issue was simply that the hardware did not meet the GIS and digitising software specifications which resulted in data acquisition and processing problems. In the latter, the use of what are now considered 'early developmental' versions of IDRISI meant that it was impossible to realise a fully operational system that made use of high topography contrast DEMs. However, it is important to note that since the initiation of the project, computer has reduced in cost in terms of initial outlay and Clark labs have now issued IDRISI version 2 and IDRISI32, both of which address the problems noted in the software section below and which are now considered industry standard applications.[38]

*Hardware*
The initial problem in this section was the lack of a powerful enough computer to process high capacity image files. Thus images and compositions took an exceptionally long time to process. More specifically movement within a map composition was hindered by a low RAM capacity, again leading to waiting times. Additionally, the physical lack of a large enough digitising tablet and configuration problems with the available A4 tablet resulted in data acquisition problems. This led to the use of a variety of digitisers, which highlighted the inadequacies of IDRISI for windows v1.01.002 (1995), specifically with compatibility of other GIS programmes and digitising software.

*Software*
IDRISI v1.01.002 (1995) proved to be inadequate for the modelling proposed, as it only accepts 8-bit real data. In order to construct high resolution DEMs, 32-bit real or 16-bit signed data are required, especially in the case of the composite DEMs which contain negative values. Additionally, poor data interchange facilities have resulted in an inability to import several data sets created on alternative digitisers, for example, the widely accepted Raw Binary format.

The digitising package (TOSCA) supplied with IDRISI was found to be inadequate for the quantity and capacity of digitising and encoding of the analogue data layers required and alternative software was used.[39] However, non-indigenous (to the system) data acquisition created problems with data format incompatibility, which resulted in the need for complicated export procedures in the digitising software (for example ARCINFO, ERDAS and PCI WORKS[40]) and import into IDRISI. Subsequently, IDRISI due to its 8-bit data capacity, was found to be inadequate for the generation of the Digital Elevation Models that included negative values and real numbers. Therefore, DEMs requiring real numbers were generated in the ORTHOENGINE Version 6.2[41] software and imported into IDRISI. Unfortunately, the import procedure converts the real numbers into integers and thus reduces the ultimate vertical output resolution. However, the preference for creating data with real values means that future work using more sophisticated GIS software will retain both a high level of accuracy and resolution.

*Data*
Increased errors associated with the use of primarily non-encoded data in analogue format datasets (i.e. sheet maps). These are published in a variety of co-ordinate systems, which means that for each dataset the co-ordinate system has to be transformed by the operator before digitisation. This is necessary otherwise the datasets cannot be used simultaneously. However, this also means that there is an increased margin of error in the final image due to accumulated errors. Additionally, archaeological projects are rarely well funded, and in that case the dataset acquisition for the construction of the GIS would probably rely heavily on manual digitisation (pre-encoded data is usually expensive), which again is a point of error introduction.

---

[38] Since the initial writing up of this work in 1999 there have been numerous developments in GIS software which would certainly address many of the problems identified.

[39] The author acknowledges the generosity of the Geography Department at Reading University, UK, for the use of the digitising facilities, and of Dr A. Ganas for technical assistance with previously unfamiliar hardware and software packages.
[40] These are the manufacturers names of alternative industry standard GIS and image processing software packages.
[41] ORTHOENGINE is a trademark of PCI GEOMATICS INC

# Chapter 6: Case Studies

Throughout this investigation a wholly interdisciplinary approach (*sensu* van Andel, 1994) to archaeoseismological appraisal has been advocated. Central to the methodology presented here are the construction of a regional empirical database (Chapter 4) and the use of a GIS system to construct seismic effects models for the site(s) under review. In this Chapter two site specific case studies from the Atalanti area were presented in order to (1) demonstrate the methodology advocated in Chapter 3, and (2) to test the working hypothesis that archaeological observations can be used as proxy indicators of seismic activity (see Chapter 3). For each site the information accrued for the regional database presented in Chapter 4 will be sifted, presented and assessed in terms of the observed data and the expected damage predicted from the GIS models in Chapter 5, and with a consideration of possible alternative mechanisms of damage. It should be noted here that despite the lack of a fully operational GIS 'effects' model (due to the absence of the digitised layers), it is not possible to produce GIS images of the possible seismic effects at the sites. However, the information collected is still valid and can be used in the site case studies in order to make the appraisal of archaeoseismology in the final chapter.

Only two site studies are presented here as broad representatives of the two main groups of topographic locations within the Atalanti region, i.e. inland and coastal sites, and the variety and breadth of archaeological sites (that are in many cases, by definition, unique) are acknowledged both in the review of the previous literature in Chapter 2 and the theoretical observations in Chapter 3. Each case study is discussed separately but follows a standard format of presentation of the evidence and discussion of the data. Additionally, because both sites lie within the same regional setting, a bullet point summary of the regional tectonic data is presented first, as a reminder of the background to the subsequent site-specific discussions. This format has been adopted specifically to demonstrate to the reader how the different attributes and contributing factors for each discrete locality affects the overall interpretation of the evidence and thus highlights why a universal identification criteria/check list approach (*sensu* Rapp, 1986, and Stiros, 1996) cannot be considered viable for archaeoseismological appraisals.

Further, by carrying out two studies representative of the two broad groupings of site locations it will be possible to make a value judgement on the usefulness of archaeological evidence for seismic activity, thus testing the working hypothesis, derived from previous interpretations, that archaeological sites can be used as proxy indicators of seismic activity. The results from this chapter will form the main conclusions to the investigation presented in Chapter 7.

## 6.1 General Background

This subsection provides a general bullet point summary of the tectonic and seismic information of the area as background to both the site-specific studies. It is purely a reiteration of the salient points, and the reader is referred to the overview presented in Chapter 4, section 4.3, for further detail.

1. Both sites lie within the extensional tectonic regime of central mainland Greece and are, therefore, influenced by large normal faults (i.e. down-throw of the hanging wall block in relation to the footwall). This results in a landscape dominated by topographic highs (trending approximately NW-SE) juxtaposed with lowland plains.

2. The seismically active Atalanti/Lokris fault is the principal tectonic structure in the region; it last ruptured in the 1894 earthquake sequence, which consisted of two main shocks estimated to be of approximately M 6.4 and M 6.9 (Ambraseys & Jackson, 1990). These events produced large scale damage throughout the plain, and beyond into the Larymna and Kammena Vourla areas, and the earthquake was felt as far afield as Athens where there was some superficial damage to the Parthenon (Martin Goalen *pers. comms.*) This Atalanti fault is also believed by several workers to be the source of the 426 BC earthquake. However, there is, at present, no unequivocal physical evidence (for example palaeoseismic trenching) to tie this event to the Atalanti fault[42].

3. All the calculations of earthquake effects used in this research are calculated assuming characteristic earthquake behaviour, i.e. that there is constant displacement per event at a given point produced by a combination of large earthquakes of a constant size and infrequent moderate events (Schwarts & Coppersmith, 1984). Additionally, the models use the maximum earthquake parameters based on those inferred from the 1894 Atalanti earthquake sequence.

4. The regional geology of the area reflects the seismic topography, with the topographic highs consisting of the pre-rift bedrock (predominantly limestone and ophiolites) in the topographic highs (footwalls) and Neogene syn-rift in the intermediate slope area, with unconsolidated quaternary (alluvial) deposits in the plain. The Neogene is predominantly Pliocene lacustrine sediments of primarily marly limestones with intermittent sandstones, thinly bedded in the region of Malesina.

---

[42] Subsequent palaeoseismic trenching has taken place on the Atalanti fault and the interested reader is directed to Pantosti *et al.,* 2004 for detailed discussion of the findings.

5. The Atalanti area is considered one of low seismicity in comparison to the rest of Greece (Papazachos & Papzachou, 1997; Stiros, 1995). Only five earthquakes (over Ms 3.9) are known (from the instrumental records) to have occurred in this area since 1964 (NOA data presented in Appendix 1 and 2) and only the 1894 event is securely known from the literature. However, it is also likely that the area experienced some ground acceleration effects from earthquakes located outside the immediate study area, for example the earthquakes of May 1893; August 1853; and 1321 whose epicentres were located in the Thebes area.

FIGURE 6.1: SITE PLAN OF THE POLYGONAL WALLS AT THE ACROPOLIS SITE AT KYPARISSI (FROM FOSSEY, 1990).

## 6.2 Inland Sites: Kyparissi

The archaeological site of Kyparissi (also known as Opous or Opus) is located 2km south-east of the hamlet of Kyparissi. It is a dual focus site, comprising an acropolis situated on the summit of Kokkinovrakhos hill (Figure, 6.1) and a possible 'settlement' site somewhere on the lower flanks of the same hill (Blegan archive material).

Fossey (1990) suggests two phases of wall construction at the acropolis, based on the observation that the style of the lower courses of polygonal walling is heavier than that of the 'truer' polygonal above. Whether this could be indicative of a destruction associated with a seismic event on the Atalanti fault cannot be ascertained. This is due to the rather vague nature of the architectural observations, which could not be corroborated by field observations (in spring 1998) as only small portions of the lower courses of the polygonal walls were still visible above the vegetation. Additionally, if the theory of two phases of construction is accepted, dating such a rebuilding event is also problematic due to the generalised nature of classification of style within polygonal walling (Shipley, *pers. comm.*). Excavation at the site for associated cultural remains (for example ceramics etc.) may remedy this problem, but until there is some corroborating evidence it is impossible to use these preliminary observations and tenuous interpretation in an archaeoseismological appraisal. For that reason, the discussion of the inland Kyparissi site focuses primarily on the lower settlement locality, where supposed archaeoseismic evidence has been previously published.

The proposed settlement site at Kyparissi (also noted Opus on the British Admiralty Map (1890), but now no longer visible) consists of a number subsurface architectural remains[43] scattered on the lower slopes of the Kokkinovrakhos hill and the farthest southeastern edge of Atalanti plain. The area was initially investigated by American archaeologists in the 1920s, and a portion of the site has subsequently been reinvestigated by the Greek Archaeological Service in 1978 and 1979. Although several trenches were opened, the published findings relate only to the base of a monumental structure found in olive groves approximately 200 metres south of the chapel of Agios Ioannis. This structure has been interpreted as the base of a stoa (a colonnaded walkway) or stylobate[44] by the principal archaeologist working at the site.

### 6.2.1 Summary of Previously Cited Evidence of Earthquake Damage

The evidence previously quoted for earthquake damage is somewhat confusing, with several publications using a combination of the same damaged structural and material cultural remains to infer a (number) of possible causative seismic events with a range of possible dates.

---

[43] It should be noted that the remains in this area must have been visible when the area was surveyed for the British Admiralty Map that was published in 1890. Two alternative explanations can be given for their subsequent absence (1) the remains have been 'robbed out' or (2) the remains have subsided in (i.e. 1894) and have been subsequently covered by sedimentation or salt marsh growth. A combination of both processes is probably the likely cause.

[44] A Stylobate is a long stone base for a row of columns, can be part of a stoa (colonnaded walkway).

*Structural*

The structural damage consists of architectural deformation in the form of upturned blockwork located in the hanging wall of a fault: see Figure 6.2. This damage has been attributed to the occurrence of an earthquake(s) on the Atalanti fault, due to the close proximity of the site to the supposed fault trace along the quaternary/bedrock contact (the geological map at Figure. 4.7 refers) (Stiros, 1988). However, it should be noted that there is no direct evidence to tie the mechanism of the observed deformation to seismic activity on the Atalanti Fault.

The date(s) of the earthquake(s) is (are) variously cited as 540 BC, 520 BC, 480 BC, 450 BC and 426 BC on the basis of archaeological ceramic evidence and parallels between the observed damage recorded after the AD 1894 event (as described by contemporary sources) with those described in the ancient texts (for the 426 BC event).

FIGURE 6.2: FIELD PHOTOGRAPH OF THE STRUCTURAL DEFORMATION (UPTURNED ASHLAR BLOCKS) OBSERVED AT THE SETTLEMENT SITE OF KYPARISSI AND CITED BY OTHER WORKERS (FOR EXAMPLE STIROS, 1988A) AS EVIDENCE OF FAULT OFFSET. (SCALE SHOWS 25CM IN 5CM DIVISIONS)

*Cultural Remains*

Two groups of terracotta roof tiles (one significantly larger than the other) are present in the site and are interpreted as being of different origins of manufacture based on style. This observation is explained by the site archaeologist as evidence of roof repair (i.e. the small group of tiles) to the main roof of the (represented by the larger group) following a destruction or damage of some kind. Although there is no specific evidence of a seismic event, the archaeologist proposes an earthquake damage hypothesis based on (1) the locality of the site, (lying, as it does, close to the seismically active Atalanti fault) and (2) the presence of structural deformation, specifically the presence of an upturned wall (see Figure. 6.2). The earthquake hypothesis is further supported using parallels between the ancient text descriptions of the effects of the 426 BC earthquake (Thucydides, Strabo, and Diodorus

Siculus; translations are given at Appendix 4) with contemporary descriptions, specifically the genesis of Donkey Island due to land subsidence as a result of the 1894 Lokis earthquake.

It is acknowledged that in addition to identification, the dating of archaeoseismic damage can be somewhat problematical. Reliable and effective dating of seismic events requires either direct evidence in the form of absolute dates (i.e. epigraphic or documentary evidence) or sound contextual associations of securely dated cultural material in association with the damaged structural/architectural remains. At Kyparissi this is not the case: Wren (1996) highlights the ambiguity surrounding the dating of the architectural terracottas (the primary evidence for constructing the occupation history of the site) and notes that the confusion over the dating of the site cannot be adequately addressed until there has been a re-examination of the terracottas. To date this has not happened.

Given the narrow time bracket of all the published dates and the fact that all the citations refer to the same excavation material (i.e. ceramics and structural remains), it is reasonable to infer that the date range is a result of archaeological dating problems rather than a series of seismic events on the Atalanti fault. This can be further supported by the fact that such a small time span does not allow for the build-up of stress required to produce a series of earthquakes large enough to result in the observed damage i.e. > Ms 4 for any surface effects (McCalpin & Nelson, 1996; Yeats *et al.*, 1997). However, damage due to event on faults outside the study area cannot be completely discounted.

In summary, five different dates are suggested (540 BC, 520 BC, 480 BC, 450 BC and 426 BC) for the destruction(s) observed in the settlement site at Kyparissi, using the same ceramic and structural evidence. From the published archaeological data the following alternative conclusions can be drawn: (1) that the site has experienced the effects of more than one earthquake occurring on the same seismogenic source, (2) that the site has been affected by more than one earthquake occurring on more than one seismogenic source either from within or outside the study area, (3) that not all the proposed dates represent an earthquake event and that some if not all may represent damage due to other natural or anthropogenic processes, or (4) that the archaeological dating is completely unreliable and no conclusions regarding the number of possible earthquake events should be drawn from the published dates.

As noted above, the dating problem cannot be resolved without re-examination of the archaeological evidence (i.e. the terracotta roof tiles). Therefore, the remainder of this appraisal focuses on the possible mechanism of the observed deformation in the structural remains.

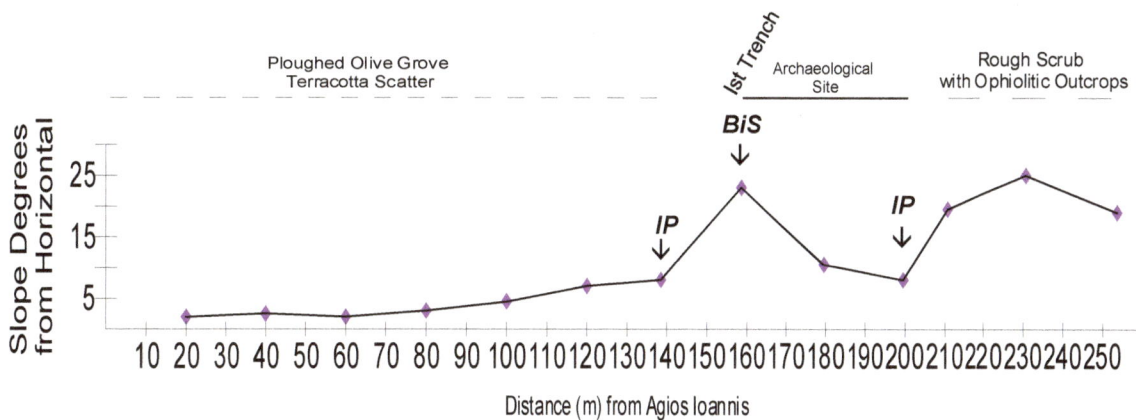

**FIGURE 6.3:** Slope profile (derived using Abney Level observations) of the alluvial fan upon which the settlement site is located. The break in Slope (BiS) is highlighted and is the location of the first trench in which the deformed foundations are located.

### 6.2.2. Geomorphological/Geotechnical Setting of the Site

The site is located within artificially terraced olive groves on a fan-like deposit on the lower slopes of the Kokkinovrakhos hill, i.e. at a slope angle range of below $10^\circ$ (Angelides, 1990, Figure 6.3 slope profile, and the GIS derived slope map at Figure A5.14)). The trenches that contain the structure lie (still open) in Quaternary unconsolidated, sand rich deposit described by Angelides (1990) as having low cohesion, on the crest of a break in slope which forms a small sub-level plateau in the gentle slope. The immediate locality is reasonably well drained with a ground water table approximately 5–10 m. below present ground level. Angelides (1990) assigns a high seismic risk factor to this deposit type, based on the geotechnical attributes.

### 6.2.3 GIS Model Prediction of the Effects at the Site

*Deformation*
Located within 1 km of the quaternary/bedrock contact in the lower south-east corner of the Atalanti plain, the site lies within the area of maximum expected subsidence, i.e. inside the 52 cm subsidence isocontour. However, due to the lack of any laterally continuous vertical reference markers (i.e. sea level) the amount of deformation cannot be quantified from the archaeological remains alone, because previous ground levels at the site are not known. The only conclusion that can be drawn is that, if a large magnitude earthquake occurs on the Atalanti fault, this site will experience a drop in elevation.

Using the overlapping Zones of Influence map (Figure 5.3), it can be seen that the elevation of the settlement site at Kyparissi would be affected by two faults: the Atalanti and Kammena Vourla faults. However, when the true deformation pattern is considered, i.e. the asymmetrical ellipses, this influence is reduced to only the Atalanti fault, as the settlement site is located on the footwall block of the Kamenna Vourla fault and is outside the

affected area. Therefore, this site cannot be used in the corroboration of supposed seismic damage observed in adjacent sites, as it does not lie within an area of overlapping ZoIs.

*Hazard*
The deterministic peak ground acceleration (PGA) calculations for Kyparissi (localities within 10 km of the focus) range from 672 gals (0.67 g) to 927.4 gals (0.93 g), calculated using values for bedrock material (see Appendix 6). However, the calculations do not take into account site effects, and given that the structure is situated on unconsolidated material (which would in effect amplify the ground acceleration (Reiter, 1990)) these numerical values can be assumed to be the minimum PGA velocities. These extremely high seismic hazard values are brought into context when considered in comparison with the 0.54 g recorded PGA in Egion for the 15/06/1995 Ms 6.2 earthquake (Koukouvelas, 1998).

The macroseismic effects derived from the intensity value (quantified using the Modified Mercelli Scale) associated with such high peak ground acceleration have been calculated to be at least VIII or IX (see Appendix 7). These figures are supported by the high seismic risk category allocated by Angelides (1990) on the basis of the geotechnical data. Within the limitations of the data, and using analogies of the scale of damage associated with modern buildings, it can be seen that at an intensity value of VIII the structure, irrespective of identity, would experience considerable damage, with perhaps the falling of roof tiles, columns and free-standing architectural decorations, e.g. sculptures. If the intensity value is increased to IX, the expected damage would increase and could include the lateral shifts of walls with respect to foundations and ground cracking in the vicinity of the surface ground ruptures. Using the more probable value of X (which takes into account localised site effects or the contribution of the geotechnical factors) it is conceivable that the building (even though constructed from monumental stonework) could experience total collapse and foundation damage. At this intensity the possibility of landslide and or slope failure in unconsolidated material of low cohesion rises significantly (McCalpin, 1996).

As noted previously, this analogy can only be used as a guide, as the scale of destruction is also a function of other associated factors, for example architectural engineering considerations, and it is unwise to assume automatic parallels between size of the earthquake (in magnitude) and any resulting damage due to the lack of data relating to the response of historical/archaeological structures to seismic excitation.

### 6.2.4 Cultural/Historical Information

*Ancient Literature*
Three ancient texts mention a fifth century BC earthquake (426 BC) in the Lokris area. Of these, one has a geographical theme (Strabo *I.3.20*), which provides details of the effects of the earthquake on a number of settlements in the Gulf of Evvia. The other two (Thucydides *iii.89* and Diodorus Sic. *XII.59.1-2*), are essentially chronological and provide very few details, but facilitate the dating of the earthquake. Collectively, all three provide little useful information with regards the observed damage at the Kyparissi site.

To date there can be no secure correlation between the texts and the actual seismogenic source of the 426 BC. The ambiguity surrounding this problem is discussed in detail in the paper by Buck and Stewart (2000). The only inferences that can be drawn from the ancient descriptions are that a large event occurred in the third century BC (with an unknown seismogenic source), and that the city of Opus in the Atalanti plain (amongst others) was seriously damaged, probably as a result of shaking associated with ground acceleration. However, this information is of limited use as the location of the city of Opus has not been securely identified, and the reference to the extent of damage is equally vague, informing the reader only 'that many sections of Alope, Cynus and Opus were seriously damaged' Strabo *(i.3.20)*. Additionally, the archaeological dating of the damage is not securely placed at 426 BC.

The city of Opus is also mentioned, by Eusebius as having been destroyed by an earthquake in AD 105/106 or 107 (depending upon the translation used). No specific details of damage are recorded, and again the lack of secure identification of the location of Opus renders this reference of little value to the Kyparissi site study, except to say that the Atalanti plain experienced ground acceleration of sufficient intensity to cause notable damage early in the first century AD.

In summary, we can conclude that from the information contained in the ancient texts there were two possible earthquakes in the Atalanti area. However, the destructions identified within the site at Kyparissi cannot be securely attributed to either due to (1) problematical identification of the location of Opus and (2) ambiguous dating of the archaeological finds (i.e. the architectural terracottas). These conclusions support the view that supposed seismic damage cannot be dated by simple cross-referencing of field observations with fragmentary literary information or by close proximity of sites with supposed seismic damage to known seismically active sources.

*Contemporary Literature*
The contemporary newspaper accounts of the 1894 Atalanti earthquake sequence described in detail the magnitude of the destruction wrought by this event. Specifically, for the area around the Kyparissi site, the following observations were made:

- Kiparissi (near Atalanti) is levelled completely
- destruction in the district south of Atalanti is even more terrible
- at Tragana, some large boulders detached from a mountain
- rock falls in the region of Kyparissi

The scientific publications indicate that the area around Kyparissi was extensively fissured; this has been variously attributed to secondary seismic land slip by Mitzopolos (1894) and to co-seismic effects by Skouphos (1894). Additionally, rock falls have been cited as having completely flattened the hamlet of Kyparissi, and having caused the movement of springs (cold water) in the Almyra area. Philippson, in his map of effects (1951) shows the site to lie in the immediate vicinity of two large fissures, one to the south and one to the north. The actual site itself is not mentioned by name.

From these reports it can be concluded that the area around the archaeological site experienced the effects of severe ground motion and slope, including rock falls and landslides. This is further supported by the geotechnical map accompanying the report by Angelides (1990), that indicates that there is evidence of serious landslides in the vicinity of the site.

### 6.2.5 Field Observations

Identification of the structure is not the aim of these field observations, which are purely for the purposes of the archaeoseismological appraisal. The structure which is cited as displaying supposed seismic damage comprises a maximum of three courses of ashlar blocks of similar dimensions (i.e. approx Length 85cm, width 42cm, and height 10cm); the largest generally having their long axis parallel to the range front quaternary/bedrock contact. The blocks, which are dressed[45] with a herringbone

---

[45] Referring to the worked surfaces.

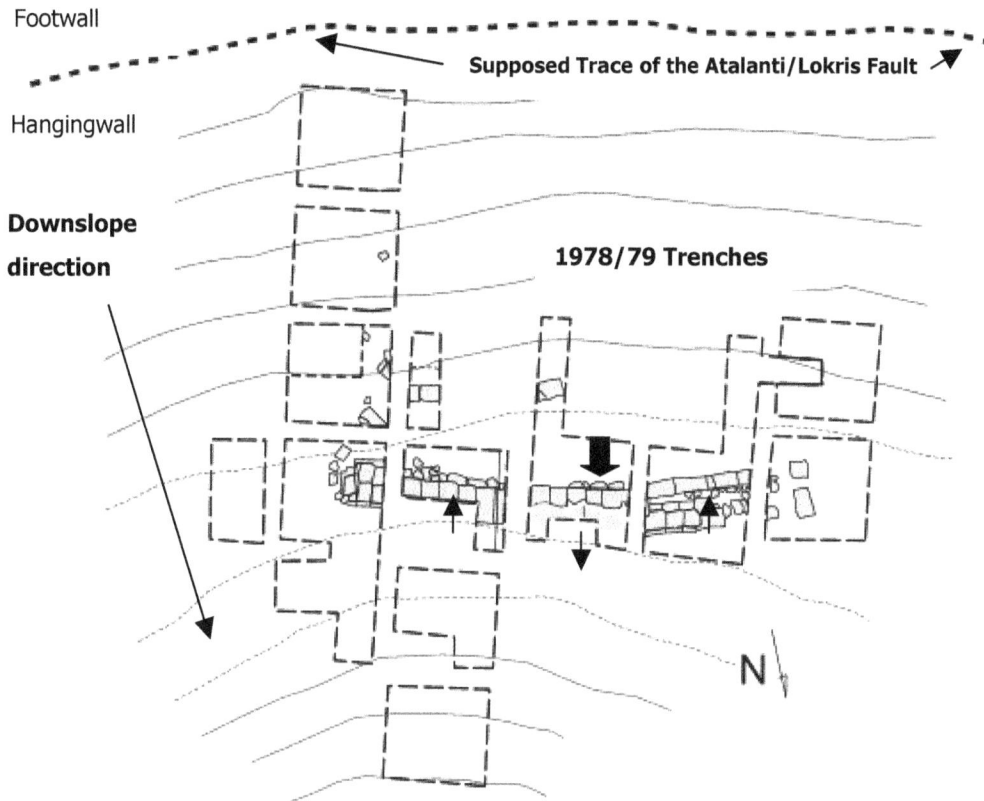

**FIGURE 6.4:** FIELD SKETCH (BASE PLAN RE-DRAWN FROM DAKORONIA 1993) OF THE PATTERN OF UNDULATION. THE BLACK ARROWS SHOW THE DIRECTION OF TILTING OF THE FOUNDATION BLOCKS: NOTE THE MIDDLE SECTION TILTS NORTH AND BOTH THE END/CORNER SECTIONS TILT SOUTH GIVING AN OVERALL APPEARANCE OF UNDULATION WITH CONSIDERABLE TWISTING.

pattern on the visible sides, are mainly cut from grey limestone with a small number cut from slightly more pinkish mylonite[46].

There does not appear to have been any bonding material (i.e. mortar), though there are possible traces of clamp and dowel[47] fixings. This feature can therefore be described as a ductile structure (Stiros 1996), and as such will accommodate an amount of movement.

The wall-like structure is not only upturned (Figure, 6.3) at the eastern end, but also twists and undulates along strike, with the central blocks generally dipping (tilting) south (inwards) towards the crest of the plateau and away from the hill (towards the north) at the ends (Figure, 6.4). Stratigraphic evidence shows the sediments abutting the blocks, with some (tenuous) evidence of back sloping layering. However, this has not been securely identified as the archaeological trench stratigraphy was unavailable.

*6.2.6 Discussion of the Possible Mechanism of Damage*

*Human Agencies*
If the 480 BC date is accepted for the destruction in the stoa, then a human agency for observable damage cannot be categorically ruled out as this is the period of unrest associated with the Persian invasions. Wren (1996) notes that it is not inconceivable that the damage was inflicted by the Persian fleet under the command of Xerxes as it advanced down the Gulf of Evvia (via Chalkis in 480 BC), possibly sacking temples and settlements en route. Although there is at present no secure evidence to unequivocally tie this hypothesis to the historical context,[48] it should be equally considered in the assessment of supposed seismic damage (Nikonov, 1988).

*Seismic Mechanism*
There are two possible seismic scenarios that could explain the observations of the conspicuous upturning and foundation undulations within the stoa. These are (1) direct offset (i.e. the upturned element of the wall) as a result of displacement on the Atalanti fault, and (2) the propagation of surface seismic waves (Rayleigh or Love waves) along strike of the fault, causing a ripple-like

---

[46] Mylonite is a fine grained gouge material.
[47] Clamps and dowels were metal fastenings used to fix the blocks in the horizontal and vertical planes respectively. These were set in poured (molten) lead in cuttings slightly larger the fastener itself. This method of fixing arrested decay by preventing air and moisture coming into contact with the iron (Camp & Dinsmoor, 1984).

---

[48] For further discussion about the possible routes taken by the Persian fleet, i.e. north or south of Evvia Island, the reader is directed to Bowen (1998).

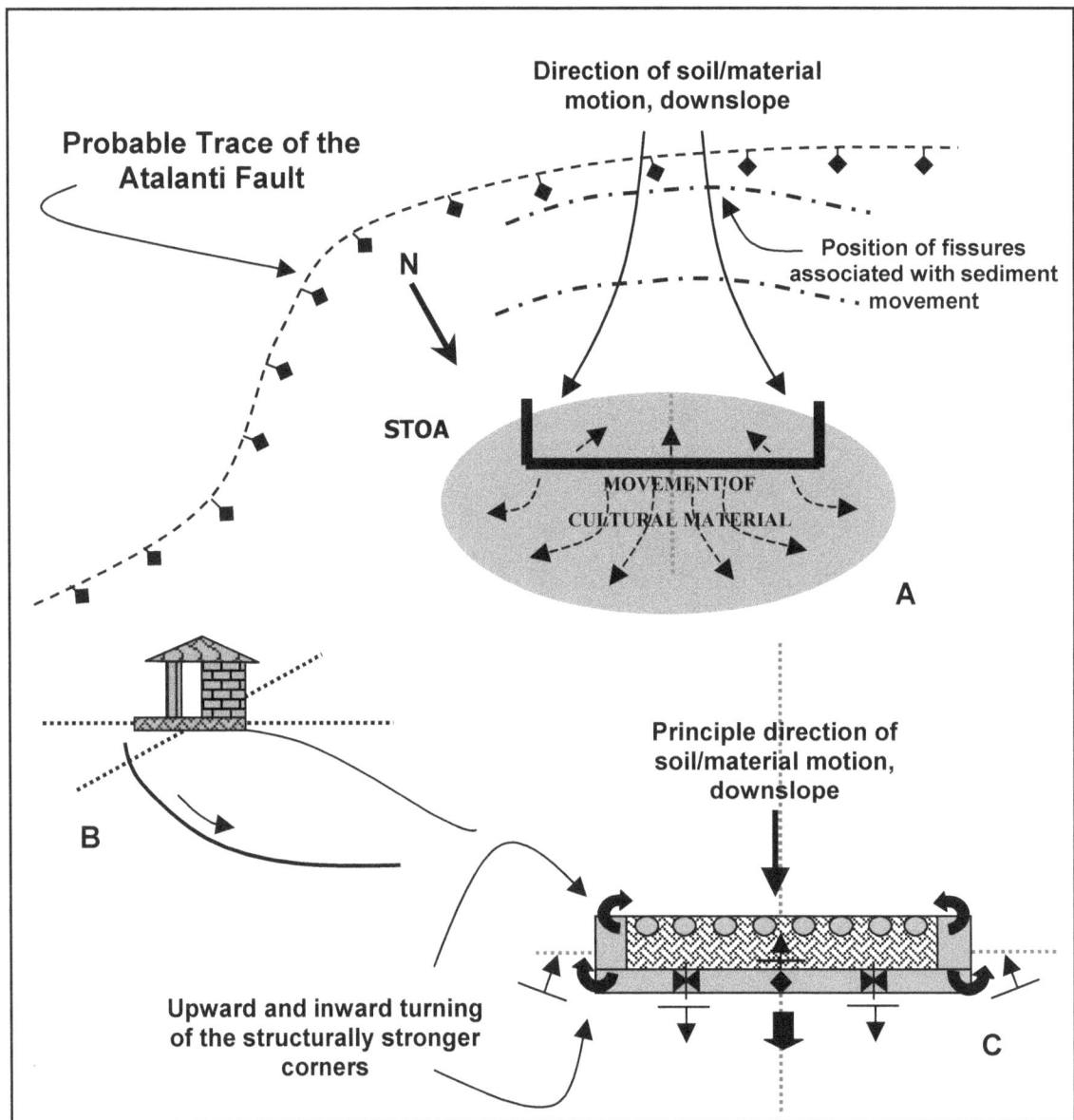

**FIGURE 6.5:** SCHEMATIC REPRESENTATION OF THE PROBABLE EFFECTS OF LANDSLIP ON THE STOA. **A,** SHOWS A PLAN VIEW, AND **B** THE CROSS SECTION AND **C** PLAN VIEW OF THE EXPECTED DEFORMATION (NOT TO SCALE)

wave action result which is preserved in the buildings as an undulating concertinaing of the stoa walls. With regards to point (1), the structure must straddle the fault trace.

However, there is no field evidence to support this theory; indeed, the long axis of the building is roughly parallel to the strike of the fault. Even invoking a small splay from the main fault trace cannot account for the inward curving of the end of the structure. Similarly, on point (2), there is at present no unequivocal physical evidence to support the fossilised seismic wave theory, and it should be highlighted that this configuration of deformation, i.e. the undulations and twisting, could also be the result of surface processes, either with or without a seismic catalyst, i.e. sediment transport down slope and/or slope failure.

*Mechanical Surface Processes.*
The observed pattern of structural deformation could also be explained as the product of ground instability processes, with the middle of the long axis of the building slipping down slope, causing the inward, or backward, tilting (towards the crest of the break in slope of the blocks) and the twisted undulations in the farthest extents of the buildings. The upturning in the east end of the structure could be explained by the pulling in of the structurally stronger corners of the building as the strike parallel (back ?) wall of the stoa slips down slope (see Figure, 6.5). Or alternatively, if the up-turning is accepted as offset, this could also be the result of displacement on a back scarp of a landslide (see Figure, 6.5). Further, it is not possible to differentiate between offset as a result of (1) primary surface sediment transport processes, or of (2) slope instability as a result of a seismic catalyst. The

inability to distinguish the mechanism from the observations is highlighted by the fact that Mitzopoulos (1894) claims that the fissuring in the Atalanti plain following the 1894 event was primarily the result of secondary surface processes, whereas Skouphos (1894) clearly states that the same fissures are seismic fractures in the quaternary deposits.

Additionally, cumulative effects of both seismic and non-seismic processes mean that it is impossible to assign a specific destructive process or event to the observed damage at this site. This can be demonstrated with the following scenario:

During earthquake A the stoa experiences severe ground acceleration (shaking) and the roof collapses (maybe only partially), resulting in broken architectural terracottas. In this event the site does not experience any ground instability. Subsequently, it is decided (for whatever reason) not to repair the razed building and the ruins are then 'robbed out' of all but the lower courses of masonry. Over a lengthy period of time the remaining courses become shallowly buried by the natural processes of site taphonomy. Some time later (say 1,500 years), earthquake B results in mass movement phenomena (due to the human intervention with the natural slope stability, i.e. terracing etc.) such as rock falls and a landslide. The sediment transport down slope includes the subsurface remains, causing buckling and twisting of the remains and the build-up of soils, etc., behind the blocks, resulting in an apparent back tilting of the sediment horizons. Again a period of time elapses and the site becomes buried and the ground surface altered by agricultural activity. Upon excavation and without reference to the context of the site, the damaged remains are, quite reasonably, interpreted as primary evidence of co-seismic destruction related to event A, and are dated using archaeological evidence and literary information, rather than a complex combination of several seismic and non-seismic processes.

Although this scenario is pure conjecture, it is presented here in order to illustrate the complexities of site taphonomy and the dangers of adopting what appears to be the simplest explanation for observed damage based on apparent correlations. It should be recalled that this site has probably been subject to the effects of more than one earthquake and that at least for the last major event (April, 1894) the area was subject to severe suficial sediment transport processes, including landslides and rock falls, which are documented in the contemporary literature, both popular press and scientific treatise.

### 6.2.7 Conclusion

The only conclusion that can be drawn from the Kyparissi site is that it would have experienced high ground acceleration (shaking) resulting from earthquakes on the Atalanti, or adjacent, faults. The response of the building to ground acceleration cannot be more clearly defined without computer modelling, but it is reasonable to infer that shaking would have been vigorous given the high intensity and PGA values. Similarly, no unequivocal conclusion regarding the dating of any seismic damage can be made, due to the present ambiguity over the dates of the architectural terracottas.

### 6.3 Coastal Sites: Alai

The archaeological site of Alai (ancient Halai) is located within the modern resort town of Theologos on the north-west coast of the Malesina peninsula. The site has been known throughout history, due to the presence of above surface remains dating from the Archaic, Hellenistic and Byzantine periods (which comprise several courses of ashlar blockwork of conglomerate material and a Christian basilica).

Subsequent excavations have revealed a long-lived site, with a wealth of subsurface Neolithic remains whose earliest levels date to c.5900 BC. The visible remains cluster around a low flat-topped hill (< 5m high) on the water's edge of the small Theologos inlet in the eastern part of the Atalanti Bay (see location map at Figure 4.3 and site plan at Figure, 6.6), approximately 10 km N of the Atalanti fault.

The site was first investigated by the Americans between 1911 and 1914 and again in 1921 and 1935 (under the directorship of Goldman and Walker), but the excavations have never been fully published. However, this deficiency is currently being addressed and the investigations updated with the more recent excavations and study seasons by the Cornell Halae and East Lokris Project (CHELP).

### 6.3.1 Summary of Previously Cited Evidence of Earthquake Damage

As with the Kyparissi site, the evidence of supposed seismic damage contained within this site is also confusing and contradictory, specifically the archaeological dating of the seismic event(s) based on correlation of architectural terracottas. Again, the recent work of Wren (1996) questions the original dating of the terracottas published by Goldman (1940), which also throws into question the dates of the supposed seismic events observed in the Kyparissi site. Goldman notes that the first temple structure was razed to the ground some time after 510 BC. The date of this catastrophe, which is identified on the basis of a thick covering of earth over damaged temple remains, is further defined and dated using references to the 426/425BC earthquake in ancient texts and correlation of the ceramic remains from the site of Herion on the Perachora peninsula (located on the Gulf of Corinth). Wren has re-dated the destruction of the first temple area to 480 BC, based on the observation that there is a distinct lack of classical material evidence. It

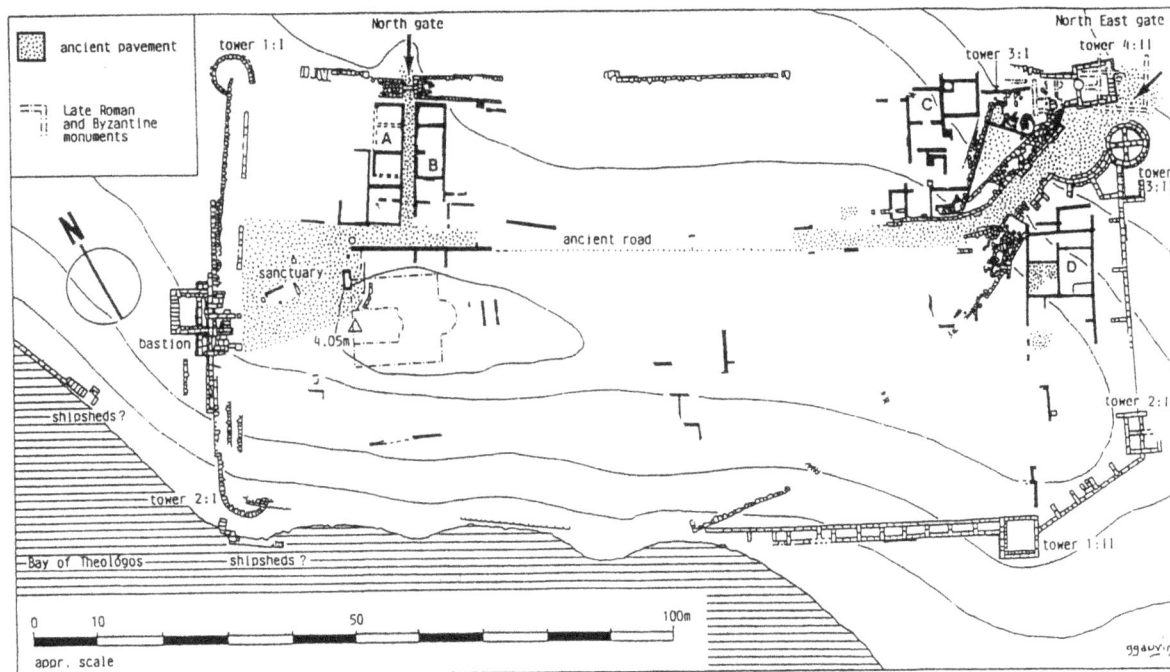

**FIGURE 6.6:** SITE PLAN OF ALAI SHOWING THE MAJOR STRUCTURAL REMAINS AS PUBLISHED BY FOSSEY 1990. NOTE THAT THE SITE EXTENDS ALL THE WAY TO THE COAST, WITH SOME REMAINS IN THE SHALLOWS. COLEMAN (1992) HAS REINTERPRETED THESE AS FORTIFICATION WALL.

should be noted, however, that Wren (1996) does not specifically allocate a seismic mechanism for the damage.

The presence of submerged relics (specifically features/walls interpreted at ship sheds)[49] at the site has led to their interpretation as evidence of a relative change in sea level since their construction, but none of the archaeological publications offers a discussion or explanation of the mechanism for these phenomena. It should be noted, however, that this site is mentioned in a geological report on the vertical movement on the Atalanti Fault by Stiros (1985), who gives a 1m estimate of the amount of subsidence, based on the presence of submerged walls.

### 6.3.2 Geomorphological Setting of the Site

The accessible part of this site (as opposed to the submerged section) is located on a Neogene (bedrock) hillock in the northern part of the small Theologos plain. The hillock itself is subhorizontal and the slope of the surrounding hillock is negligible, less than 2 degrees. Considerable reconstruction and beach stability measures (i.e. concrete retaining walls and promenade) round the full extent of the Theologos bay divide the beach from the rest of the plain and from the intertidal and littoral sections, but again both these areas have a very shallow

slope angle of less than 2 degrees (from the GIS DEM of the Theologos bay).

The fortification walls embrace the whole hillock and thus the main archaeological remains. Excavation has revealed that the earliest levels (i.e. the early Neolithic) of the Alai are founded directly on the Neogene bedrock. Structures outside the walls are founded on quaternary deposits consisting of unconsolidated terrestrial deposits in the eastern part with overlying beach deposits (i.e. pebbles and beach rock) to the seaward side. Angelides (1990) does not include this area in his geotechnical study; therefore a more detailed break down of the plain deposits is not available. Within the fortification walls and to the north-east and east the unconsolidated material is well drained, whilst to the seaward sections the ground water table, which is barely subsurface, is influenced by the tidal fluctuations. That the sea level has changed since the earliest habitation of the site is evident because:

1. there are a variety of structures (mainly walls) located offshore and in the littoral zone, which overlain by beach rock containing terracotta in poor condition, and
2. the lowest levels of the habitation remains within the fortification walls on the hillock are tidally waterlogged by fluctuating ground water levels. At high tide, brackish water stands some several centimetres in the trenches, well above the lowest Neolithic foundation materials.

---

[49] The identification of the features/walls is not secure illustrated by the presence of question marks next to the interpretations on the plans included in p.37 of Fossey (1990).

### 6.3.3 GIS Model Prediction of the Effects at the Site

*Deformation*

Using the modified deformation ellipses from Ganas (1998) and Ganas and Buck (1998) the expected deformation at Alai, associated with a single large magnitude earthquake on the Atalanti fault, would be 30cm. Unlike the Kyparissi site there is the potential to observe this deformation due to the presence of a laterally extensive vertical marker, i.e. sea level. However, it may not be possible to identify deformation associated with a single earthquake event, as the expected subsidence (i.e. approx. 30cm) is less than the tidal range, observed to be as high as 1m in some areas of the bay.

*Hazard*

To date there are no published probabilistic seismic hazard calculations for the Alai site. However, given (1) the close proximity of the site to the two sites where values are available (Ganas, 1998), (2) the similarities in the parameters used in the probabilistic calculations, and (3) the fact that these are essentially regional scale calculations (i.e. 1:50,000) it is reasonable to suggest here that the hazard values would be in the range of 0.05 g (i.e. Kynos) and 0.06 g (Kyparissi) for a 37 % exceedence of PGA in a 1,000-year recurrence interval. This is presents a low hazard, as would be expected for an area that has experienced a large magnitude event with the last 150 years.

The deterministic PGA calculated for this site, based on bedrock medium and 100% probability of occurrence results in extremely high, ranging between 0.53 g and 0.63 g. This would cause major ground motion (shaking), with felt intensity values of between VIII and IX. The similarity between the values at the sites is due to the similar $R^{50}$ values (i.e. approx. 10 km) as a result of the inclined plane of the Atalanti fault, which dips north under the Alai site. This in effect gives only a 4 km difference in location of the sites from the hypocentre (see Appendix 6), even though the difference in surface distance of the sites from the fault trace is nearer to 10 km. These high ground acceleration and intensity values could result in damage similar to those described for the Kyparissi site, including cracking in walls, lateral shifts of walls on foundations, the falling of roofing material, columns and architectural decorations and possibly the total collapse of structures. However, unlike Kyparissi there would be no possibility of damage due to land instability, due to the lack of a slope and the fact that the structures are founded on bedrock.

As noted above, there are no published geotechnical data for quaternary deposits of the Malesina penninsula. Based on the fact that the hillock (on which the majority of the archaeological remains are located) comprises Neogene bedrock, it is reasonable to extrapolate a seismic risk allocation on the basis of similar bedrock composition to the Neogene on the opposite side of the bay at the Xerorouni plateau (see Figure 4.3, location map). However, this coarse correlation cannot be employed for the unconsolidated sediments of the plain, as they would require specific analysis to derive such factors as cohesion, drainage, etc. Therefore no geotechnically derived seismic risk allocation can be assigned for this locality. It should also be noted that it is not necessary to consider these intensity values as the minimum, as this site is founded upon bedrock and therefore there will be no geotechnical contribution.

### 6.3.4 Cultural/Historical Information

*Ancient Literature*

This site is not mentioned in any of the ancient sources in relation to seismic damage. It is, however, mentioned a number of times, in relation to the sacking of the town by the Roman general Sulla in AD 85 when it was razed to the ground, and in Pausanias' travels in Greece in the second century AD (hence its identification by the antiquarians and early travellers).

*Contemporary Sources*

Again Theologos (or the archaeological site of Alai) is not specifically mentioned in any of the contemporary literature pertaining to the 1894 earthquake sequence. This is unsurprising, as until fairly recently, when the area was significantly developed for tourism and holiday homes, the settlement was a very small fishing hamlet, essentially an offshoot of the larger site of Malesina, with few residents. This lack of references in the contemporary literature highlights a major problem with using historical account of pre-instrumental seismic events to derive any earthquake parameters; namely that the reports rely heavily on population distribution and density. Where small and/or remote villages sustain severe effects, but are relatively unknown and unvisited by literate persons, the effects go completely unknown and earthquake is potentially lost in history. Conversely, where a relatively small event affects a major population point, i.e. a town or city, the effects can be disproportionately described and a minor earthquake can become a major catastrophe (Ambraseys, 1971; Ambraseys, 1983). This in turn will influence any seismic hazard assessments that use recurrence interval as one of the calculation parameters.

The scientific publications give no data for this area. Although Philippson (1894) maps extensively the remains of the surface features for the 1894 event, he indicates no phenomena on the peninsula. Similarly Skouphos shows no effects at this area, and indeed even the inundation recorded in the Atalanti plain, which must surely have affected the Theologos plain, also goes unrecorded in his text. However, Papavassilore (1894) notes that a wave submerged the full extent of the coast and specifically records St Theologos as the one of the farthest extents of its presence.

---

[50] The distance from the hypocentre to the site.

In summary, there are no ancient references to earthquake effects in the site of Alai and only scant reference to a sea wave associated with the 1894 event as having affected the hamlet of Saint Theologos. There are no details of the damage, if any, sustained at the archaeological site in any of the publications currently known.

### 6.3.5 Field Observations

There were no observations within the site of what could be perceived as damage associated with a seismic event. Some possible observations from previous excavation seasons were pointed out by the excavation director (Prof. Coleman); these included a single (small) stretch of a toppled habitation wall (Neolithic), two laterally extensive pale fine-grained sediment deposits (indicative of sea wave incursion?) and the previously mentioned thick deposit of red sediments, similar to pounded red clay, in the remaining unexcavated first temple area[51] There were no indications of foundation disturbance or shifts (i.e. opening of jointings or displacement of blocks in relation to each other) in the later Archaic or Hellenistic remains, i.e. in the fortification walls or in the subterranean burial chambers present on the site. This can be considered noteworthy given the deterministic PGA expected at this site.

### 6.3.6 Discussion of the Mechanisms of Damage

#### Human agencies

If the original dating of the damage by Goldman is correct, then there is no clear evidence to suggest human actions as the possible mechanism of destruction. However, if the archaeological evidence is re-dated to 480 BC (as suggested by Wren, 1996), then again this would coincide with the advance of the Persian fleet (under the command of Xerxes) down the Gulf of Evvia towards Athens via the channel at Chalkis. Given the invasive nature of the fleet, it is not unreasonable to suggest that the Persians could have plundered the settlements along the shores of the Evvian Gulf (including the coastal settlement of Alai) en route. Whereas this scenario is purely conjectural, since there is no material or literary evidence to support this hypothesis, the fact that the fleet would have had to make daily stops (for provisions and to rest the oarsmen), coupled with the location Alai as a coastal settlement means that the possibility of a human mechanism for the observed damage cannot be categorically ruled out. Moreover, it is not unreasonable to suggest that it is a reasonable inference that the lack of any unequivocal evidence of damage at this time could be explained by immediate rebuilding and re habitation following (1) the advance of the Persians or (2) the historically recorded sacking (and razing to the ground) of the town by the general Sulla in AD 85.

#### Earthquake

There are no unambiguous published data to suggest earthquake destruction at this site. However, the presence of the red sediment deposit, discussed by Goldman and observed by the author, may be attributable to major collapse and subsequent pounding carried out in order to level the mudbrick walls that usually constituted the upper parts of buildings, especially habitations in the Archaic period. This seems increasingly more plausible when the general physiography of the site is taken into consideration, and it is noted that there is no obvious natural source or cause for the clay deposit that was initially observed by Goldman (1940). As there is no recorded information of cultural inclusions, i.e. ceramics or other material remains, it is not possible to ascertain without a doubt the deposit origin or formation process. It may be that sediment analysis would reveal the original use, if any, of the clay.

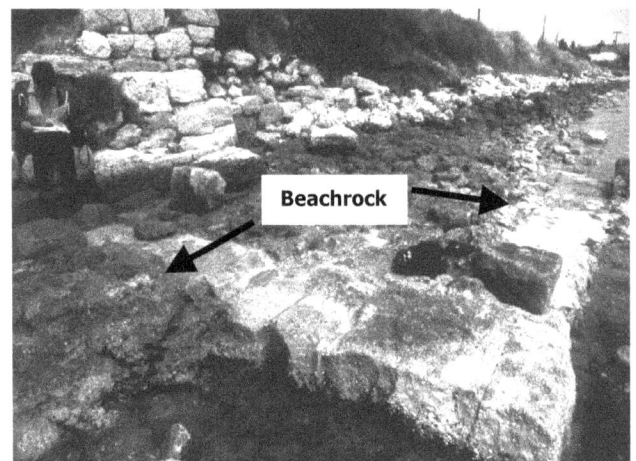

**FIGURE 6.7:** FORTIFICATION FOUNDATION WALLS OF THE SITE OF ALAI IN THE INTERTIDAL ZONE WITH OVERLYING BEACHROCK DEPOSITS. NOTE THAT THE BEACHROCK HAS BEEN REMOVED FROM THE WALLS DURING INVESTIGATIONS BY THE GREEK ARCHAEOLOGICAL SERVICE, UNDERWATER DIVISION

#### Sea-Level Changes

There are clearly two indisputable lines of evidence for an apparent change in sea level that can be derived from the archaeological remains. Both the following observations are further supported by the presence of terracotta tiles and ceramics as constituents of the beach rock which overlie the foundation walls observed in the water (Figure 6.7).

1. The earliest Neolithic levels of the habitation area are waterlogged at high tide, with the water rising through the Neogene marls.

2. The presence of anthropogenic structures (walls) underwater. Although these were originally thought to be shipsheds, they have re-interpreted by Coleman (*pers. comm.*) as foundations of buildings and part of the fortification walls. Additionally, an underwater survey by the Greek

---

[51] The extent of the deposit is undeterminable as it has been mainly removed (all but the surviving small section in the first temple area) during the previous excavations.

Archaeological Service (Underwater Division) has shown that the remains extend farther into the bay than originally thought[52] again questioning the shipshed hypothesis.

Although the submerged walls in the second point could be reinterpreted again in the future, it can be concluded without doubt that there has been a relative change in sea level since the earliest occupations as it is highly unlikely that the Neolithic inhabitants would construct their dwellings where they would get wet feet twice a day. This interpretation is further supported by the presence of graves (currently interpreted as from the Mycenaean period) below the current sea level on Mitrou Island. Clearly, the graves would not have been on or below sea level when originally constructed; therefore, they support the hypothesis of a significant change in relative sea level, although dating such an event is problematical. This means that the minimum subsidence can be estimated using the archaeological vertical datum and the levels for the earliest remains.

There are two possible mechanisms for the apparent relative change in sea level: (1) a eustatic[53] change in sea level and (2) neotectonic (seismic) activity, i.e. movement on the Atalanti fault due to earthquakes. Lambeck (1995) has calculated the eustatic contribution to sea level change over an extended period. It can be seen that there was a major rise of sea level around 6000 BP of around 48 m[54] in the region of the Atalanti Bay. This rise may well explain the occupation highs observed in the archaeological stratigraphy after the end of the early Neolithic period when perhaps the site was no longer viable due to the loss of the potential grazing or the agricultural hinterland that would have extended over the whole of the Atalanti Bay. This inference is derived from the fact that the current bathymetry shows the water depth as being up to 50 m[55] which would clearly have been above the contemporary sea level if the current level is dropped by 48 m. It is worth noting here that such a large difference is sea level would have rendered the physiography of the whole of the Atalanti plain area very different and therefore any interpretation of the earliest Neolithic landscape should ideally take into account pre-seismic topography (in the form of pre seismic DEMs, see Chapter 5).

The tectonic contribution to the post Archaic (i.e. after approx. 2500 BP) change in sea level can be calculated using true deformation ellipses associated with known

seismogenic sources in the area. This method was used by Ganas and Buck (1998) to calculate a likely co-seismic contribution of 60 cm from two earthquakes on the Atalanti fault and a hypothetical offshore (Malesina Peninsula) fault. Additionally, a 45 cm post-seismic contribution was calculated (on the basis of Ma & Kusznir, 1995), giving approximately 1.06 m of tectonic subsidence over a 4000-year period, i.e. from the Archaic period, with negligible eustatic contribution[56]. These calculated values are in accordance with archaeological data for submergence at the site of Alai. Again the precise dating of the events that produced the observed subsidence is problematic, and only a 'ball park' figure, i.e. two events within 4,000 years, can be assigned.

### 6.3.7 Conclusion

In conclusion, even though the deterministic ground acceleration values show that the site could experience up to intensity ten, there is no unequivocal evidence of ground motion (or shaking) at this site, even for the 1894 Atalanti earthquake sequence. However, the presence of a laterally continuous vertical datum means that it is possible to observe the effects of seismic deformation. However, problems still exist, including (1) differentiation between co-seismic and post-seismic deformation and (2) quantification of the subsidence in terms of specific measurements. As with the Kyparissi site, the dating of seismic events from archaeological observations is problematic as it requires the presence of archaeologically datable material in good contextual association with geological phenomena.

### 6.4 Discussion

The archaeoseismological appraisals carried out for this chapter aimed to test the working hypothesis that observations of seismic events in archaeological remains can be used as proxy indicators for earthquake geology and palaeoseismology and, in doing so confront the challenge of Charles Richter: 'that ancient accounts of earthquakes [either as texts or physical remains] do not help us much; they are incomplete, and accuracy is usually sacrificed to make the most of a good story' (Richter, 1958, cited in Vita-Finzi, 1986. p 8). The conclusions of this work show that different sites (both archaeological, historical and contemporary) will yield a range of earthquake information both in accuracy and resolution, thus highlighting the role of variable contributing factors and the problems of adopting universal criteria or all encompassing methodologies which do not allow for a full interdisciplinary appraisal of the available data.

---

[52] Information derived from the CHELP archive held at the excavation centre in Theologos, with the permission of the excavation director, Prof. John Coleman.

[53] Eustaic relates to the global change in sea level as appose the localised change attributable to neotectonic activity.

[54] Not necessarily in one single instantaneous rise, it was probably more likely that the rise took place in stages over a short period of time which is unidentifiable in the geological record.

[55] The 50m contour runs across the Atalanti Bay from the Malesina peninsula toward Kynos, approx 2km seaward of Atalanti Island, see Figure 4.3.

---

[56] Eustatic contributions from Lambeck (1995)

# Chapter 7: General Conclusions

The research aimed to investigate the subdiscipline of archaeoseismology (focusing on the extensional tectonics of central Greece) in terms of theory, methodology and practical application. The results of this work can be summarised in the following paragraphs:

The extensive literature review in earlier chapters has shown that there are two major problems associated with using archaeological observations of seismic damage. These are (a) that the evidence used by archaeologists and seismologists was diverse and cross-disciplinary, and (b) that the evidence is highly subjective in terms of interpretation of observations. Although methodological papers advocate multidisciplinary research as a means to address points (a) and (b), this was rarely practised. Moreover, interpretations (and subsequent citations of supposed seismic damage) a more commonly made on the basis of identification criteria or a check list approach to field observation with scant regard to both the human and the physical context of the site.

The detailed investigation into the reasons for this apparent methodological stagnation has highlighted the previous lack of understanding of the problems associated with data from unfamiliar disciplines, i.e. the theoretical, methodological and practical considerations. This is especially the case where the data under consideration are derived from such wide-ranging, and clearly disparate, subject areas, for example archaeology, history, seismicity, geotechnics and coastal research. Additionally, the limitations of the collected data are rarely considered within the context of the actual aims of the research, for example the resolution of the dating and location of events.

A new methodology, based on the construction of a project-specific research strategy, has been proposed here in order to address the methodological problems which have been highlighted. The primary difference between this and previous methodologies is that it advocates a truly interdisciplinary, approach where data (in a range of scales) are collected by workers in their own specialist area and brought together in a regional database by a co-ordinator. The product of such an approach is the construction of a regional database which can be used as a data source for a range of projects, present and future.

A Geographical Information System has been proposed as a tool to assist in the construction of the regional database, thereby addressing the logistical problems associated with collation of such a large number of diverse datasets. Additionally, the GIS was proposed as a means of deriving information useful for archaeoseismological appraisal from the apparently obscure data. For example, the construction of zone of influence and neutral effects models were constructed for

the two sites, and were used in this thesis to make an evaluation of previously cited observations of damage interpreted as seismic in origin.

An important advantage of using a GIS for the storage, manipulation and analysis of the data required for archaeoseismological appraisal is that additional information can be added when it becomes available, and the database is thus a dynamic repository which can be used as a primary source for a diversity of projects, both in terms of both mono- and inter-disciplinary projects; for example, traditional excavation interpretations, archaeoseismological appraisal and palaeo-environmental reconstructions.

## 7.1 In Summary

The primary goal was to evaluate the usefulness of archaeologically derived seismic information as a source of proxy data for seismological research. The application of the alternative methodology proposed in this thesis allowed for a more balanced assessment, and showed that

- different archaeological observations can be considered more reliable indicators than others. For example, squashed skeletons in non-ritual burial environments with associated damage can be assigned to a seismic mechanisms more readily than deformed walls or toppled columns. This leads to the conclusion that

- each archaeological site will have variable potential to provide information for different aspects of a seismic event, due to the diversity of contributing factors, both natural and anthropogenic, and related to the relics themselves and their physiographic setting. This point alone highlights the deficiencies of an approach that uses a universal check list of identification, and supports the individualised site appraisal suggested in the theory and methodology chapter. Additionally,

- many of the observations used to infer a seismic event could equally be the products of secondary seismic effects, e.g. landslides or mechanical processes such as soil creep, and/or of anthropogenic processes, e.g. wilful destruction. Additionally, where a sesimic mechanism has been securely allocated, it is still extremely difficult to discern whether the observations are from individual or multiple events, or from the effects of a combination of seismic and non-seismic factors. And finally,

- the resolution of the information that can be gained from archaeological observations does not always match the resolution necessary for seismic research. For example, the amount of co-seismic deformation gained from submerged remains will depend on knowing the original absolute position of structures in relation to the previous sea level. This point emphasises the need to interpret archaeoloseismological observations within a broader regional context, and warns against a monodisciplinary approach.

## 7.2 Future Research

A major deficiency with the research carried out here is that that it has not been possible to achieve a fully integrated GIS system as discussed in Chapter 3. As such many of the results contained within this publication are preliminary, for example the zones of influence are derived using the symmetrical distance operator rather than true deformation ellipses. As noted in the Chapter 5, the problems lie in the lack of digital data for the area, the lack of data transformation hardware, and the limited data handling capacity of version one of the IDRISI GIS software. Therefore, in the first instance a future aim is to construct a fully integrated GIS system, using a software package that has the capacity to deal with the large data array associate with large topographic contrasts and the ability to construct interactive text/tabulated data boxes to ensure the full integration of the textual and numerical data. Additionally, the method outlined in Chapter 3 of this thesis has only been used in an extensional regime. A further field test of the methodology should be carried out in a separate tectonic regime to ascertain whether the methodology is transferable, or is limited by the circumstances of its construction, i.e. only suitable for active normal faulting. Coupled with this aspect is the necessity to gain a broader experience as a whole in the type of archaeological observations that are, or have been, interpreted by archaeologists as indicative of seismic events. However, this in itself throws up a separate area of future research which has yet to be addressed, and which has only been touched upon in this thesis. That is the subjectivity within the identification and interpretation and the method for quantification of field observations in such a way as to make archaeologically derived earthquake information useful for other disciplines, i.e. more than stratigraphic benchmarkers in archaeological interpretations.

Two possible methods require further investigation

1. a reliability index for the observations of damage, which can be incorporated in
2. the use of logic tree formalisation to provide numerical values as probabilities of the uncertainties attached to the field observations that can be incorporated in seismic hazard investigations.

# Appendices

# Appendix 1

**A1.1:Table of Earthquake Data Supplied by the National Observatory in Athens (November 1996).**

Earthquakes recorded by the National Observatory in Athens, Greece, from July 1964 to November 1996 between Lat 38.20 - 38.50 and Lon 22.30 - 23.30, over Mag 3. The highlighted events (i.e. Mag = 4.0 and above) are used in the EQ layer of the Atalanti GIS.
*(Data supplied by the National Observatory of Athens, Nov 1996 in ASCII formatt).*

| Date | | | Time | | | Lat | Long | Depth | Mag |
|------|------|----|------|----|------|------|------|-------|-----|
| **1964** | **JUL** | **12** | **23** | **32** | **50.5** | **38.50** | **23.25** | **5** | **4.9** |
| 1965 | JUL | 16 | 13 | 54 | 27.7 | 38.50 | 23.25 | 5 | 3.8 |
| 1965 | OCT | 19 | 17 | 08 | 29.5 | 38.80 | 22.70 | 5 | 3.8 |
| **1965** | **OCT** | **28** | **04** | **27** | **10.5** | **38.40** | **22.50** | **5** | **4.6** |
| 1965 | NOV | 10 | 20 | 25 | 8.4 | 38.50 | 23.25 | 5 | 3.6 |
| 1966 | JUL | 21 | 04 | 11 | 8.2 | 38.50 | 23.10 | 5 | 3.5 |
| 1966 | SEP | 2 | 06 | 47 | 47.8 | 38.50 | 23.00 | 5 | 3.6 |
| 1966 | SEP | 7 | 13 | 00 | 28.5 | 38.40 | 23.30 | 5 | 3.6 |
| 1967 | JAN | 10 | 10 | 35 | 24.1 | 38.40 | 23.10 | 5 | 3.5 |
| 1967 | JUN | 11 | 13 | 31 | 15.0 | 38.50 | 22.75 | 5 | 3.6 |
| 1967 | AUG | 3 | 01 | 19 | 58.0 | 38.40 | 22.60 | 5 | 3.4 |
| 1967 | AUG | 25 | 13 | 15 | 8.0 | 38.50 | 23.00 | 5 | 3.6 |
| 1967 | NOV | 3 | 09 | 46 | 1.0 | 38.40 | 22.90 | 5 | 3.3 |
| **1968** | **JAN** | **21** | **13** | **58** | **38.0** | **38.40** | **23.20** | **5** | **4.2** |
| 1968 | DEC | 10 | 20 | 13 | 23.0 | 38.50 | 23.00 | 5 | 3.8 |
| 1969 | JAN | 29 | 09 | 38 | 36.0 | 38.50 | 22.70 | 5 | 3.5 |
| 1969 | MAR | 25 | 04 | 18 | 8.0 | 38.40 | 22.60 | 5 | 3.4 |
| 1969 | MAR | 25 | 04 | 37 | 2.0 | 38.50 | 23.00 | 5 | 3.2 |
| 1969 | MAR | 25 | 05 | 57 | 9.0 | 38.50 | 22.50 | 5 | 3.9 |
| 1969 | MAR | 25 | 06 | 10 | 33.0 | 38.50 | 22.50 | 5 | 3.3 |
| 1969 | AUG | 26 | 04 | 59 | 25.0 | 38.50 | 22.50 | 5 | 3.2 |
| **1969** | **SEP** | **24** | **08** | **26** | **34.0** | **38.40** | **22.60** | **5** | **4.2** |
| **1969** | **OCT** | **2** | **23** | **13** | **37.0** | **38.50** | **22.40** | **5** | **4.4** |
| **1969** | **NOV** | **22** | **20** | **34** | **40.0** | **38.40** | **23.30** | **5** | **4.3** |
| 1969 | NOV | 22 | 20 | 38 | 46.0 | 38.40 | 23.20 | 5 | 3.1 |
| 1969 | NOV | 23 | 06 | 39 | 14.0 | 38.50 | 23.30 | 5 | 3.3 |
| 1969 | NOV | 23 | 10 | 33 | 33.0 | 38.40 | 23.10 | 5 | 3.3 |
| 1969 | NOV | 23 | 16 | 59 | 15.0 | 38.40 | 23.20 | 5 | 3.3 |
| 1969 | NOV | 26 | 01 | 25 | 51.0 | 38.40 | 23.20 | 5 | 3.1 |
| 1970 | FEB | 21 | 23 | 59 | 46.0 | 38.40 | 23.10 | 5 | 3.3 |
| 1970 | MAR | 1 | 19 | 39 | 33.0 | 38.40 | 22.40 | 5 | 3.4 |
| 1970 | APR | 8 | 15 | 10 | 57.0 | 38.40 | 22.60 | 5 | 3.5 |
| 1970 | APR | 8 | 15 | 56 | .0 | 38.50 | 22.70 | 5 | 3.7 |
| 1970 | APR | 8 | 16 | 38 | 47.0 | 38.40 | 22.70 | 5 | 3.5 |
| 1970 | APR | 8 | 19 | 30 | 32.0 | 38.50 | 22.70 | 5 | 3.8 |
| 1970 | APR | 8 | 21 | 30 | 13.0 | 38.40 | 22.70 | 5 | 3.8 |
| 1970 | APR | 8 | 22 | 16 | 57.0 | 38.40 | 22.70 | 5 | 3.6 |
| 1970 | APR | 9 | 01 | 07 | 27.0 | 38.50 | 22.60 | 5 | 3.7 |
| 1970 | APR | 9 | 09 | 43 | 25.0 | 38.40 | 22.60 | 5 | 3.3 |
| 1970 | APR | 9 | 20 | 50 | 19.0 | 38.40 | 22.70 | 5 | 3.7 |
| 1970 | APR | 9 | 22 | 44 | 21.0 | 38.90 | 22.90 | 5 | 3.2 |
| 1970 | APR | 11 | 15 | 56 | 22.0 | 38.50 | 22.80 | 5 | 3.4 |
| 1970 | APR | 13 | 07 | 41 | 36.0 | 38.50 | 22.60 | 5 | 3.3 |
| 1970 | APR | 14 | 08 | 59 | 31.0 | 38.40 | 22.80 | 5 | 3.2 |
| 1970 | APR | 15 | 22 | 14 | 20.0 | 38.40 | 22.70 | 5 | 3.3 |
| 1970 | APR | 16 | 19 | 48 | 19.0 | 38.40 | 22.60 | 5 | 3.4 |
| 1970 | APR | 20 | 16 | 21 | 9.0 | 38.50 | 22.75 | 5 | 3.4 |
| 1970 | APR | 27 | 21 | 11 | 34.0 | 38.50 | 22.80 | 5 | 3.3 |
| 1970 | MAY | 8 | 00 | 16 | 45.0 | 38.40 | 22.60 | 5 | 3.9 |
| 1970 | MAY | 13 | 09 | 24 | 20.0 | 38.40 | 22.80 | 5 | 3.1 |
| 1970 | MAY | 27 | 18 | 35 | 34.0 | 38.50 | 22.70 | 5 | 3.3 |
| 1970 | JUN | 29 | 05 | 11 | 27.0 | 38.50 | 23.00 | 5 | 3.2 |
| 1970 | AUG | 5 | 23 | 27 | 2.0 | 38.50 | 23.30 | 5 | 3.2 |
| 1971 | JAN | 5 | 05 | 33 | 9.0 | 38.40 | 22.90 | 5 | 3.5 |
| 1971 | FEB | 27 | 08 | 20 | 43.0 | 38.40 | 23.00 | 5 | 3.6 |

```
1971 MAR 13   17 17  4.0   38.40   22.90   5        3.5
1971 MAR 14   00 44 30.0   38.40   22.90   5        3.6
1971 APR  3   10 52 48.0   38.40   22.90   5        3.2
1971 JUN 16   23 14  6.0   38.40   22.90   5        3.3
1971 JUN 23   06 51 25.0   38.40   22.40   5        3.5
1971 JUL  1   07 10 47.0   38.50   23.10   5        3.2
1971 NOV  8   00 52  5.0   38.50   23.00   5        3.5
1971 NOV  8   18 48 21.0   38.40   22.90   5        3.5
1972 MAR 22   22 07 28.0   38.40   23.10   5        3.0
1972 MAR 22   22 09 55.0   38.40   23.20   5        3.1
1972 MAR 22   23 51 34.0   38.40   23.20   5        3.2
1972 NOV 25   15 20 46.0   38.50   22.40   5        4.2
1973 JAN 21   21 36 57.0   38.50   23.30   5        3.3
1973 JUN 14   01 57 10.0   38.50   22.50   5        3.5
1974 JAN  5   08 45 15.0   38.40   22.60   5        3.2
1974 APR 13   05 15 27.0   38.40   22.70   5        3.3
1974 JUL  6   16 02 47.0   38.50   22.75   5        3.2
1974 AUG  4   02 16  7.0   38.40   22.40   5        3.3
1974 AUG  7   12 47 39.0   38.50   23.10   5        3.0
1974 NOV 14   12 41 29.0   38.40   23.10   5        4.3
1974 NOV 14   12 51 18.0   38.40   23.00   5        3.2
1974 NOV 14   13 22 32.0   38.40   23.10   5        5.2
1974 NOV 14   13 34 43.0   38.50   23.10   5        3.5
1974 NOV 14   14 05 48.0   38.40   23.10   5        3.6
1974 NOV 14   14 22 10.0   38.50   23.20   5        4.0
1974 NOV 14   14 26 49.0   38.40   23.10   5        5.2
1974 NOV 14   14 31 47.0   38.40   23.00   5        4.4
1974 NOV 14   14 43 49.0   38.50   23.10   5        4.0
1974 NOV 14   14 50 33.0   38.50   23.10   5        4.3
1974 NOV 14   14 55 29.0   38.50   23.00   5        4.1
1974 NOV 14   15 05 43.0   38.50   23.10   5        3.4
1974 NOV 14   15 29 44.0   38.50   23.00   5        5.3
1974 NOV 14   18 33 18.0   38.50   23.10   5        4.1
1974 NOV 14   22 11 19.0   38.40   23.00   5        3.6
1974 NOV 14   23 11 13.0   38.40   23.00   5        3.6
1974 NOV 15   04 00 49.0   38.40   23.10   5        4.1
1974 NOV 15   08 14 36.0   38.50   23.10   5        3.4
1974 NOV 15   09 11 51.0   38.40   23.10   5        3.2
1974 NOV 15   10 51  8.0   38.50   23.10   5        3.6
1974 NOV 15   10 51 27.0   38.40   23.10   5        3.9
1974 NOV 15   15 37 47.0   38.40   23.00   5        3.6
1974 NOV 15   16 17 58.0   38.40   23.10   5        3.5
1974 NOV 16   02 45 31.0   38.40   23.10   5        3.1
1974 NOV 16   05 19 40.0   38.50   23.10   5        3.6
1974 NOV 16   13 28 28.0   38.50   23.00   5        4.1
1974 NOV 16   19 38 16.0   38.50   23.10   5        3.3
1974 NOV 17   00 02 10.0   38.50   23.10   5        4.2
1974 NOV 18   01 12 26.0   38.50   23.10   5        3.4
1974 NOV 20   20 34  8.0   38.50   23.00   5        3.2
1974 NOV 21   02 55 38.0   38.50   23.00   5        3.2
1974 NOV 21   14 05 41.0   38.50   23.00   5        4.0
1974 NOV 21   15 15  9.0   38.40   23.00   5        3.1
1974 NOV 21   18 05 48.0   38.50   22.90   5        3.6
1974 NOV 23   09 20 44.0   38.50   23.20   5        4.3
1974 NOV 23   12 45 50.0   38.50   23.25   5        3.3
1974 NOV 24   03 12 58.0   38.50   23.20   5        4.1
1974 NOV 24   03 25 43.0   38.50   23.10   5        3.3
1974 NOV 25   16 09 35.0   38.50   23.10   5        3.3
1974 NOV 26   04 24 45.0   38.40   23.20   5        3.0
1974 NOV 27   21 51 40.0   38.50   23.20   5        4.1
1974 DEC  1   05 13 59.0   38.40   23.00   5        4.3
1974 DEC  1   06 21 13.0   38.40   23.00   5        4.9
1974 DEC  1   06 26 11.0   38.50   22.90   5        3.3
1974 DEC  1   06 27 42.0   38.50   22.90   5        3.5
1974 DEC  1   06 34  7.0   38.50   22.90   5        3.4
1974 DEC  1   11 43 20.0   38.50   22.90   5        3.3
1974 DEC  2   12 49 18.0   38.40   23.00   5        3.1
1974 DEC  2   13 49 59.0   38.40   23.00   5        3.2
1974 DEC  2   14 07 24.0   38.40   23.00   5        3.4
1974 DEC  4   23 21 35.0   38.50   23.00   5        4.5
```

```
1974 DEC  5   14 54 54.0   38.50   23.10    5        3.5
1974 DEC  7   20 06 15.0   38.50   23.10    5        3.2
1974 DEC  8   00 09 18.0   38.50   23.10    5        3.8
1974 DEC  8   11 12  3.0   38.50   23.30    5        3.1
1974 DEC  8   17 27 10.0   38.50   23.10    5        3.1
1974 DEC  9   02 16 34.0   38.50   23.40    5        3.0
1974 DEC  9   06 55 22.0   38.50   23.20    5        3.1
1974 DEC 10   01 22 15.0   38.50   23.10    5        3.6
1974 DEC 10   20 16 11.0   38.50   23.00    5        3.2
1974 DEC 11   07 58 27.0   38.40   23.00    5        3.0
1974 DEC 11   10 40 50.0   38.40   23.10    5        3.1
1974 DEC 13   08 06   .0   38.40   23.00    5        3.2
1974 DEC 16   01 01  2.0   38.50   23.20    5        3.4
1975 JAN  1   10 45 42.0   38.40   22.80    5        4.2
1975 JAN  1   10 54 28.0   38.40   22.80    5        3.6
1975 JAN  1   11 14 30.0   38.40   22.80    5        3.3
1975 JAN  8   19 28 12.0   38.40   23.00    5        4.3
1975 JAN  8   20 16 18.0   38.50   22.70    5        3.2
1975 JAN  8   20 29 59.0   38.40   22.70    5        3.3
1975 JAN  9   01 16  5.0   38.50   23.10    5        3.2
1975 JAN  9   05 39 28.0   38.40   22.80    5        3.3
1975 JAN 10   04 58 47.0   38.50   23.00    5        3.2
1975 JAN 10   05 31  3.0   38.50   23.00    5        3.8
1975 JAN 12   13 57 24.0   38.40   23.00    5        4.3
1975 JAN 18   00 45 16.0   38.40   22.80    5        3.6
1975 JAN 19   02 29 57.0   38.50   23.00    5        3.2
1975 JAN 19   15 44 31.0   38.50   23.00    5        3.2
1975 JAN 20   16 42 36.0   38.50   23.00    5        3.2
1975 JAN 23   14 47  1.0   38.50   23.25    5        3.1
1975 FEB  3   21 56 37.0   38.40   22.80    5        3.4
1975 FEB 17   12 44 25.0   38.50   23.30    5        3.5
1975 MAR  2   16 15 26.0   38.40   23.00    5        3.2
1975 APR  1   08 20  2.0   38.50   23.20    5        4.5
1975 APR  1   08 38 40.0   38.50   23.10    5        3.6
1975 APR  3   15 56 31.0   38.40   22.40    5        3.5
1975 APR 10   06 59  4.0   38.40   23.10    5        3.4
1975 MAY 13   00 31 10.0   38.40   22.50    5        3.4
1975 MAY 13   22 42 52.0   38.40   22.70    5        3.2
1975 JUN  5   01 57 44.0   38.50   23.00    5        3.2
1975 OCT 12   10 30 11.0   38.50   22.80    5        3.5
1975 DEC  9   15 40 18.0   38.40   22.40    5        3.4
1976 JAN 11   15 00 10.0   38.50   22.60    5        3.6
1976 FEB  4   16 09 14.0   38.50   23.00    5        3.3
1976 MAR  4   23 13 25.0   38.40   23.10    5        3.7
1976 MAR 24   19 00 29.0   38.40   22.40    5        3.3
1977 AUG 12   08 44 30.4   38.40   22.50   10        3.5
1977 SEP  2   13 58  9.0   38.40   22.80    5        3.0
1977 SEP  5   00 29 51.0   38.50   22.90    5        3.5
1977 SEP  7   09 45 38.0   38.40   22.80    5        3.9
1977 SEP 18   14 48 27.0   38.40   22.90    5        3.3
1977 OCT  6   17 12 14.0   38.50   22.90    5        3.2
1978 FEB 15   19 08 19.0   38.40   22.40    5        3.4
1978 APR 28   05 44 24.0   38.50   23.10    5        3.2
1978 SEP  9   16 32   .0   38.40   23.10    5        4.9
1978 SEP 29   19 45  7.0   38.40   22.40    5        3.5
1979 JAN 16   11 57 32.0   38.50   23.30    5        3.9
1979 JAN 22   05 49 42.0   38.50   23.00    5        3.5
1979 JUN  2   17 53 37.0   38.40   22.40    5        3.8
1979 JUN  8   20 47 39.0   38.50   23.00    5        4.4
1979 JUN  9   11 06  4.0   38.50   22.90    5        3.9
1979 JUN 25   02 32 40.0   38.50   23.00    5        3.6
1979 JUL 29   22 33 29.0   38.40   23.00    5        3.5
1980 JAN 31   04 20 37.0   38.50   22.60    5        3.6
1980 FEB 29   17 15 44.0   38.40   23.10    5        3.3
1980 MAR 11   02 41 31.0   38.40   23.10    5        3.2
1980 MAR 31   10 57 47.0   38.40   23.10    5        3.0
1980 MAY 15   20 36 59.0   38.40   23.10    5        3.4
1980 JUN 16   00 25 43.0   38.40   23.20    5        3.2
1980 JUL  7   07 57 41.0   38.40   22.90    5        3.6
1981 FEB 26   10 26 13.0   38.40   23.30    5        3.8
```

```
1981 FEB 26    11 23 26.0    38.40    23.00     5      3.9
1981 FEB 26    12 21 56.0    38.40    23.20     5      3.7
1981 FEB 26    13 01 20.0    38.40    23.20     5      4.0
1981 FEB 27    06 54 53.0    38.40    23.30     5      4.4
1981 FEB 27    15 32 36.0    38.40    23.30     5      4.0
1981 FEB 28    12 23 30.0    38.40    23.20     5      3.8
1981 MAR  6    02 05 14.0    38.50    23.60     5      4.1
1981 MAR  9    20 25 39.0    38.50    23.10     5      4.3
1981 MAR 10    02 09 24.0    38.40    23.10     5      3.7
1981 MAR 11    03 40    .0    38.40    23.10     5      4.2
1981 MAR 14    09 52 35.0    38.40    23.30     5      3.7
1981 MAR 17    21 52 50.0    38.40    23.20     5      4.1
1981 MAR 19    13 28  7.0    38.40    23.30     5      3.8
1981 MAR 21    06 45 47.0    38.40    23.30     5      3.9
1981 MAR 22    10 18 39.0    38.40    23.30     5      3.7
1981 MAR 22    10 47 48.0    38.40    23.30     5      4.0
1981 MAR 23    01 07 35.0    38.40    23.30     5      4.0
1981 MAR 24    11 35 52.0    38.40    23.30     5      4.4
1981 APR  2    00 47 41.0    38.40    23.30     5      3.7
1981 APR  9    14 19 51.0    38.40    23.10     5      3.7
1981 APR 18    08 07  2.0    38.50    23.10     5      4.8
1981 APR 20    18 10 15.0    38.40    23.00     5      3.6
1981 APR 28    07 19 59.0    38.40    22.40     5      4.1
1981 MAY 14    00 28 45.0    38.50    22.50     5      3.6
1981 JUN 28    07 31 56.0    38.40    23.00     5      3.7
1981 JUN 29    22 02  6.0    38.40    23.20     5      4.3
1981 JUL 14    07 17 28.0    38.40    23.20     5      4.4
1981 AUG  4    19 49 11.0    38.50    23.00     5      3.5
1981 AUG  5    01 59 43.0    38.40    23.30     5      3.7
1981 AUG  9    05 22 25.0    38.40    23.20     5      3.6
1981 AUG 17    00 44 12.0    38.40    23.20     5      3.8
1981 SEP 27    01 33 34.0    38.50    22.50     5      3.4
1981 OCT  1    21 43    .0    38.40    22.80     5      3.8
1981 DEC 22    22 36 23.0    38.40    22.70     5      3.8
1981 DEC 26    09 01    .0    38.40    22.70     5      3.7
1982 JAN 12    08 32 41.0    38.40    23.20     5      3.4
1982 FEB 18    06 22 24.0    38.40    22.90     5      3.3
1982 FEB 23    03 19  8.0    38.50    22.90     5      3.3
1982 APR  5    20 28 49.0    38.40    23.30     5      3.5
1982 MAY 18    18 18  8.0    38.40    23.30     5      3.1
1982 JUL 28    22 13 35.0    38.40    23.00     5      3.5
1982 AUG  1    20 31 52.0    38.40    23.00     5      3.3
1982 AUG  3    04 46 14.0    38.40    23.20     5      3.0
1982 AUG  5    19 21 47.0    38.50    23.10     5      3.7
1982 AUG  9    14 03 20.0    38.40    23.10     5      3.4
1982 AUG 12    15 52 48.0    38.40    23.00     5      3.8
1982 AUG 12    15 59 21.0    38.50    23.10     5      3.1
1982 AUG 23    21 13 18.0    38.40    23.10     5      3.2
1982 AUG 24    18 30    .0    38.50    23.00     5      3.2
1982 AUG 25    01 26 26.0    38.40    23.00     5      3.1
1982 SEP 29    12 02  2.0    38.50    23.10     5      3.2
1982 OCT 20    12 10 34.0    38.50    22.40     5      3.8
1982 NOV 20    01 38 13.3    38.43    23.24    10      3.2
1982 NOV 28    08 04 36.2    38.49    23.02    37      3.4
1983 AUG 21    19 11 42.8    38.50    23.28    27      3.0
1989 NOV 26    19 12 21.9    38.45    23.02     5      3.1
1992 MAR 28    18 49 29.4    38.40    22.49    32      3.2
1992 NOV 10    13 47 42.4    38.43    22.45    36      3.5
1992 NOV 11    16 15 23.1    38.44    22.49    45      3.6
1993 FEB  4    11 19 42.8    38.44    22.78    98      3.5
1995 JUN 15    05 01 38.7    38.49    22.70    37      3.7
1996 MAR  3    17 31  2.5    38.42    22.63     5      3.3
1996 JUN 12    11 57 25.9    38.48    22.60    35      3.3
```

# Appendix 2

**Table A2.1**: Extracted Earthquakes (>Mag 3.9) from July 1964 to November 1996 between Lat. 38.20 - 38.50 and Long. 22.30 - 23.30, and used in the EQ layer of the Atalanti GIS.
(*Data supplied by the National Observatory of Athens, Nov 1996 in ASCII formatt*)

| Date YY MM DD | Time HH MM SEC | LONG | LAT | DEPTH Km | MAG |
|---|---|---|---|---|---|
| 1964 JUL 12 | 23 32 50.5 | 38.50 | 23.25 | 5 | 4.9 |
| 1965 OCT 28 | 04 27 10.5 | 38.40 | 22.50 | 5 | 4.6 |
| 1968 JAN 21 | 13 58 38.0 | 38.40 | 23.20 | 5 | 4.2 |
| 1969 SEP 24 | 08 26 34.0 | 38.40 | 22.60 | 5 | 4.2 |
| 1969 OCT 02 | 23 13 37.0 | 38.50 | 22.40 | 5 | 4.4 |
| 1969 NOV 22 | 20 34 40.0 | 38.40 | 23.30 | 5 | 4.3 |
| 1972 NOV 25 | 15 20 46.0 | 38.50 | 22.40 | 5 | 4.2 |
| 1974 NOV 14 | 12 41 29.0 | 38.40 | 23.10 | 5 | 4.3 |
| 1974 NOV 14 | 14 22 10.0 | 38.50 | 23.20 | 5 | 4.0 |
| 1974 NOV 14 | 14 31 47.0 | 38.40 | 23.00 | 5 | 4.4 |
| 1974 NOV 14 | 14 43 49.0 | 38.50 | 23.10 | 5 | 4.0 |
| 1974 NOV 14 | 14 50 33.0 | 38.50 | 23.10 | 5 | 4.3 |
| 1974 NOV 14 | 14 55 29.0 | 38.50 | 23.00 | 5 | 4.1 |
| 1974 NOV 14 | 15 29 44.0 | 38.50 | 23.00 | 5 | 5.3 |
| 1974 NOV 14 | 18 33 18.0 | 38.50 | 23.10 | 5 | 4.1 |
| 1974 NOV 15 | 04 00 49.0 | 38.40 | 23.10 | 5 | 4.1 |
| 1974 NOV 16 | 13 28 28.0 | 38.50 | 23.00 | 5 | 4.1 |
| 1974 NOV 17 | 00 02 10.0 | 38.50 | 23.10 | 5 | 4.2 |
| 1974 NOV 21 | 14 05 41.0 | 38.50 | 23.00 | 5 | 4.0 |
| 1974 NOV 23 | 09 20 44.0 | 38.50 | 23.20 | 5 | 4.3 |
| 1974 NOV 24 | 03 12 58.0 | 38.50 | 23.20 | 5 | 4.1 |
| 1974 NOV 27 | 21 51 40.0 | 38.50 | 23.20 | 5 | 4.1 |
| 1974 DEC 1 | 05 13 59.0 | 38.40 | 23.00 | 5 | 4.3 |
| 1974 DEC 1 | 06 21 13.0 | 38.40 | 23.00 | 5 | 4.9 |
| 1974 DEC 4 | 23 21 35.0 | 38.50 | 23.00 | 5 | 4.5 |
| 1975 JAN 1 | 10 45 42.0 | 38.40 | 22.80 | 5 | 4.2 |
| 1975 JAN 8 | 19 28 12.0 | 38.40 | 23.00 | 5 | 4.3 |
| 1975 JAN 12 | 13 57 24.0 | 38.40 | 23.00 | 5 | 4.3 |
| 1975 APR 1 | 08 20 2.0 | 38.50 | 23.20 | 5 | 4.5 |
| 1978 SEP 9 | 16 32 .0 | 38.40 | 23.10 | 5 | 4.9 |
| 1979 JUN 8 | 20 47 39.0 | 38.50 | 23.00 | 5 | 4.4 |
| 1981 FEB 26 | 13 01 20.0 | 38.40 | 23.20 | 5 | 4.0 |
| 1981 FEB 27 | 06 54 53.0 | 38.40 | 23.30 | 5 | 4.4 |
| 1981 FEB 27 | 15 32 36.0 | 38.40 | 23.30 | 5 | 4.0 |
| 1981 MAR 6 | 02 05 14.0 | 38.50 | 23.60 | 5 | 4.1 |
| 1981 MAR 9 | 20 25 39.0 | 38.50 | 23.10 | 5 | 4.3 |
| 1981 MAR 11 | 03 40 .0 | 38.40 | 23.10 | 5 | 4.2 |
| 1981 MAR 17 | 21 52 50.0 | 38.40 | 23.20 | 5 | 4.1 |
| 1981 MAR 22 | 10 47 48.0 | 38.40 | 23.30 | 5 | 4.0 |
| 1981 MAR 23 | 01 07 35.0 | 38.40 | 23.30 | 5 | 4.0 |
| 1981 MAR 24 | 11 35 52.0 | 38.40 | 23.30 | 5 | 4.4 |
| 1981 APR 18 | 08 07 2.0 | 38.50 | 23.10 | 5 | 4.8 |
| 1981 APR 28 | 07 19 59.0 | 38.40 | 22.40 | 5 | 4.1 |
| 1981 JUN 29 | 22 02 6.0 | 38.40 | 23.20 | 5 | 4.3 |
| 1981 JUL 14 | 07 17 28.0 | 38.40 | 23.20 | 5 | 4.4 |

# Appendix 3

**Table A3.1: Information transcribed from Contemporary Sources for the Earthquakes of 1894**

| Date | Source | Remarks |
|---|---|---|
| 21.04.94 | Times | A strong shock was felt in Athens at 6 55 pm.<br>Direction of shock was apparently West - East, duration half a minute with increasing intensity |
| 23.04.94 | Times | Eight villages in the district of Atalanti totally destroyed<br>Pier at the harbour at Pelli was rent asunder and sank beneath the sea<br>At Tragana, some large boulders detached from a mountain.<br>At Atalanti, greatly damaged, houses fell, chasms opened in the neighbourhood<br>Thebes affected, Chalkis affected<br>Violent shocks felt throughout Euboea, and neighbouring islands of Skiathos, Skopelos, Skyros<br>Shocks felt in Athens |
| 24.04.94 | Times | Shocks continue throughout yesterday and today [23$^{rd}$] and houses previously damaged now fallen<br>Shocks somewhat less violent during the last 24hrs<br>An almost incessant trembling of the earth is still perceptible here<br>A considerable amount of damage in Athens and Piraeus<br>Parthenon sustained injuries |
| 26.04.94 | Times | (Correspondent in the district of Atalanti)<br>Kato Pelli, row of houses, which forms the only street, completely destroyed.<br>End of the pier has sunk beneath the sea, the remainder of the structure thrown to one side and will soon go to pieces.<br>Destruction wrought in the district south of Atalanti is even more terrible<br>The shocks of the last 24hrs have been comparatively slight, |
| 27.04.94 | Times | (Correspondent in the district of Martini [Martinon].)<br>Everywhere scene on terrible destruction<br>Kiparissia (near Atalanti) levelled completely<br>Proskyna completely destroyed<br>Malesina, completely levelled to the ground<br>Martini, suffered equally (correspondent writes from here) |
| 28.04.94 | Times | (Correspondent in the district of Martini)<br>Severe shock felt at 10:30 (27$^{th}$) Great damage was done at Atalanti, the ground being torn up.<br>Villages, Topolinas severely damaged. Kardinitsa almost completely escaped. |
| 30.04.94 | Times | A severe shock on Friday night, Atalanti was at the centre of the disturbances. A fissure opened in the ground and extended in a SW direction to a distance of eight or ten km.<br>The site of the town subsided to a depth of one meter and a half.<br>The seismic wave swept over the shore of Pelli.<br>The neighbouring village and monastery of St. Constantine were destroyed…Great stones falling from the mountains rolled over Kyparissia and other villages. Loud reverberations accompany the shocks and succeed one another rapidly. At Limnie and Xerochorion, in Euboea, the earthquake was extremely violent.<br>Observatory (Athens) report that 12 shocks in the last 48hrs |
| 01.05.94 | Times | At many places hot springs of considerable volume have appeared; at others the wells have dried up. Shocks of earthquake, accompanied by subterranean noises have continued during the last 24hrs |
| 02.05.94 | Times | Inroads of sea extended inland for 3km<br>Chasm in earth @ 12km extends from town<br>Surface of sea (in many places) milky aspect due to submarine discharges of sulphurous matter<br>A strong shock felt [that day] throughout Athens. It was especially severe in Euboea. |
| 04.05.94 | Times | Earthquakes continue to be felt one of unusual duration.<br>Boiling springs of great volume appeared at Aedipsos, disclosing the remains of the Roman Baths |
| 07.05.94 | Times | Another severe shock of earthquake occurred at 5am in the districts of Livadia, Thebes and Atalanti |
| 09.05.98 | Levant & Eastern Herald | Issued by Reuters Agency. Eight whole days the country has been shaken from end to end with 365 shocks in the space of 7.5hrs. Sea encroached some 18 – 20 m on land due to a large seismic sea wave. Huge masses of rock have fallen down from the mountains<br>Rumblings from the depths of the earth<br>The shore on the Euboean side has sunk 2 meters<br>At the sulphur springs of Aidipsos torrents of hot water issued forth, and fresh springs appeared |
| 10.05.94 | Times | A circular fissure has appeared in the ground enclosing the village of Xarma and surrounding district. The space enclosed shows symptoms of subsiding |
| 12.05.94 | Times | Heavy rain and hail in Atalanti |

| 17.05.94 | Times | Several shocks have been experienced in the last few days. Three fissures extend through the site [of Atalanti] Chasm near Skandagara is 250m L, 50 B, 15 D <br> Hot springs at Aedipsos leaping like geysers with temp at boiling point Rain fell in torrents |
|---|---|---|

## Table A3.2: Information on the 1894 earthquakes derived from Science Sources

| Date | Source | Author | Observations |
|---|---|---|---|
| 03.05.94 | Nature Vol 50 N°. 1279, p7 | C Davidson | Severe earthquake in Greece, April 27 at 9:20 <br> Observed on a bipendular instrument in Birmingham <br> Details of timings and measurements using the instruments. |
| 03.05.94 | Nature Vol 50 N°. 1279 p3 | Anon | Severe disturbance, centre in Atalanti. <br> Times correspondent reports that the inroads of the sea in this district extend for a distance of 3m <br> Surface of the sea in many places is coloured with the products of submarine eruptions <br> Chasm has opened in the ground not far from Atalanti and extends in a SW direction for 12km |
| 18.10.94 | Nature Vol 50, N° 1303 | C Davidson | Abstract of the translations of Papvasiliore by Davidson <br> Earthquake series consisted of two principal shocks and many minor ones <br> 1st occurred on **20 April** and was registered of the seismoscope at Athens <br> The region which experienced much damage can be split into three zones according to their damage <br> **Epicentral** *(villages completely destroyed)* :comprising the peninsular of Aetolymion, <br> **Zone Two** *(nearly all the buildings overthrown)*: ellipse whose major axis = 28km extending SE - NW from Larymna Bay to Cape Arkitza, minor axis 8-9km <br> **Zone Three** *(many houses were damaged or partially fell)* ellipse, major axis 90km directed SE-NW reaching from Dritza to near Molos, minor axis 65km extending from Levadia to Mantoudi (Evvia). <br> • The night of the 20-21, almost constant disturbance, interrupted by often stronger shocks <br> • Frequent shocks for 3 days throughout all three zones <br> 2nd violent shock occurred at 9pm on **27 April** followed by the same continual ground motion. <br> **Zone Two:** major axis increased by 30km, especially towards the NW. From Scroponeri [Larymna] to St Constantin. <br> **Zone Three:** major axis increased by 22km to the town of Lamia. <br> • Minor axis of both zones increased several km to the SW. <br> • A sea wave rose up with the 2nd shock and submerged the whole Bay from St. Theologos to St Constantin. The water retired except in the Bay of Atalanti where the greater part of the coast is now submerged for a distance of some meters. <br> • Springs ceased to run, whilst others increased their flow. <br> • New Thermal springs at Aedipsos near pre-existing ones. <br> • Numerous fissures, some kilometres in length, have been formed…the most remarkable is the great 55km fissure. Breadth varies a few centimetres to 3m, but on average 1.5m. Extends from Bay of Scroponeri through Atalante in a constant ESE-WNW direction <br> • The fissure is identified as a fault because of <br>   1. Extrodinary length and parallelism to the Gulf of Evvia <br>   2. Consistancy of direction and independence of geological structure <br>   3. Existance of throw (generally small, 0cm on Cretaceous to cm's on Tertiary and 1.5m on Alluvial beds) and horizontal displacement causing a lowering of the Atalanti plain and a slight shift to the NW <br> • Papavasiliore regards this as one of the faults that gave rise to the Gulf of Evvia at the end of the Tertiary/beginning of Quaternary. |

| 1895 | pp 29-32 | Mitzopoulou | The spring area SE of Kyparissi was separated from the sea by a narrow strip of land (no dimensions given). The main surface break is explicitly stated as commencing from the channel of Karagiozi (NW of Atalanti) and is described as 10 - 15 km and terminating at Almyra. The town of Atalanti became uninhabited. The break direction is not clear, but the primary break formed an 'enveloping' surface to the Tertiary beds of the Atalanti plain. *This suggests that the break may have followed the pre-rift syn rift contact.* A narrow strip of land (10-20m) parallel to the coastline at, Skala-Palaeomagaza, subsided and was covered by the sea. *subsidence only listed a having occurred south of Skala!* Donkey Island became an island The village of Kato Pelloi (old Skala which may be slightly landward) water from the ground aquifer changed colour (dissolved clays sand etc.). Also one of the wells in this village was rotated CCW 9(0?) to the NW The breaks were visible from the sea, (but he does not say which breaks) The Break (we assume main break) is positioned upslope of the old city of Atalanti. Pazari is specifically mentioned as one of the localities (this site we visited on the 1997 field season, it is totally covered by debris, so not even the old architecture is visible… but old men in the café were asked where the breaks had been and they pointed to the exact spot in the Skuphos paper) Smaller breaks were present downslope of Atalanti He specifically denies the existence of breaks in the Malesina peninsula (W of Larymna) Mitzopoulou basically denies that the breaks represent a seismogenic fault. He claims landsliding of the alluvium material that covers the Atalanti plain. Breaks were mapped outside the Atalanti plain, to the NW, at areas called 'Xarma' and Achladi' Near(east) of Rengini, now deserted. |
|---|---|---|---|

**Table A3.3: Topographic and Travel Writings that give information on the pre 1894 Geography/Topography**

| DATE | AUTHOR | TITLE | VOL/PAGE | | OBSERVATIONS |
|---|---|---|---|---|---|
| 1819 | Dodwell E | Classical and Topographic Tour through Greece | 2 | 57 | At the pass of Andera, remains of a strong wall about 70 paces from the foot of a steep hill to the margin of the sea. A stream issues from the rock and enters the sea [*Almyra area*] |
| | | | | | Ancient remains (foundations) ¼hr from the pass |
| | | | | | Specifically notes "near the Locrian shore there is an island" then quotes Diod.Sic, Strabo & Pliny for the formation of the island (see Lolling 1876 for the argument against this view) and mentions that this coast is very liable to earthquakes |
| | | | | | 1hr from Andera, remains of a city, the soil 'which is considerably raised above its ancient level'… which is probably the site of Opus. |
| | | | | | In a description of the gulf Dodwell states "the smooth expanse of water is broken by the island of Atalanti, and some long promontories which shoot out from the Locrian shore" |
| 1819 | Gell W | The Itinerary of Greece | - | 228 - 233 | Notes a gate and walls of massive stones that defend the pass between the mountains and the sea. |
| | | | | | Many salt and fresh water founts are noted. A small islet is described with 'vestiges' . |
| | | | | | The tower [Frankish] is noted on the Xlomon mountain. A large village of 'Alachi' is noted on the coast with salt works |
| | | | | | A 'bed of a torrent' before reaching a church on left at foot of Mt Talanta with a marble chair and other vestiges [before Atalanti]. |

| 1835 | Leake W M | Travels in Northern Greece | 2 | 184-185 | Discusses the possibility of the Copiac Lake draining through the mountains and emerges in the springs by the Almyra area Map shows Donkey Island attached to the mainland coast. |
|------|-----------|----------------------------|---|---------|------|
| 1872 | Many | Handbook for travellers in Greece, 4<sup>th</sup> edition | | 229 | Map included shows two islands in Atalanti Plain, (pre 1894!!) Donkey Island is clearly detached. 'Proskymno is 5hrs [from Architza]; passing by the scala of Talanda, leaving the twon itself 2m to the rt. Talanda derives its name from the islet of Atalanta, which shelters in its port. Remains of the ancient city of Opus are found at Kardenitza, a village 1hr S.E' |
| 1873 | Tozer H F | Lectures on the Geography of Greece | - | - | Map drawn by Dr C Muller at 1:1,333,333 shows the [Donkey] Island peninsula connected to the main coastline |
| 1916 | Oldfather W A | Studies in the History and Topography of | - | 155 | |
| 1965 | Liddell R | Mainland Greece | - | 33 | Alatanti port of some consequence in the Middle Ages |
| 1985 | Pritchett W K | Studies in Ancient Greek Topography | | | Mentions three earthquakes but mainly in ref to Oldfathers (1939) work. EQ's are dated 426BC, AD 106, April 1894. |
| 1911 | Johnson, Sanborn | Private letter to Hill of the American School in Athens | | 3 | 'A great stretch of the shore beyond the town [Kyparissi] is partly submerged, and changes along the coast at this point in very recent times are apparent: Gaidrouronesi, a peninsula when Bursian visited the locality, about 1860, is now an island, [adeed in pencil] since the great earhquake of 1894' |
| 1918 | Hydrographic Office | The Mediterranean Pilot | | 230 | General charts 426, 2836b. Chart 1556, Gulf of Volo, Oreos and Talanta Channels [*read with copy of chart*] <br><br> Atalanti Bay: Western Side of Bay (*Sinus Opuntius)* two islands, Atalanti and Gaidaro, the latter is nearly united to the shore by low marshy land and forms the north west side of a bight about 1¾miles deep called Port Armyro which has 7 - 3 fathoms of water. A rock with 7 feet water upon it lies nearly in the middle of the entrance, which is about ½ mile wide. <br><br> Atalanti Island, northward of Gaidaro, [height etc] and its southern end is only seperated from the shore by a narrow passage 5 - 7 fathoms deep. |

# Appendix 4: Ancient Texts

### Thucydides, ii. 32. and iii.89.1-4

"Towards the end of this summer (431 BC) the Athenians also fortified and garrisoned Atalante, the island which lies off Opuntian Lokris and had hitherto been unoccupied. Their object was to prevent pirates sailing from Opus and the other ports of Lokris and ravaging Euboea. These were the events that took place during this summer after the withdrawal of the Peloponnesians from Attica"

(Thycydides,Lo.i.314 - 315)

"In the following summer the Peloponnesians and their allies, led by Agis son of Archidamus, king of the Lacedaemonians, advanced as far as the Isthmus with the intention of invading Attica; but whenmany earthquakes occurred, they turned back again, and no invasion took place. At about the same time, while the earthquakes prevailed, the sea at Orobiae in Euboea receded from what was then the shoreline, and then coming on in a wave overran a portion of the city. One part of the flood subsided, but another engulfed the shore, so that what was land before is now sea; and it destroyed of the as many people as could not run up to the high ground in time. In the neighbourhood also of the island of Atalante, which lies off the coast of Opuntian Locris, there was a similar inundation, which carried away a part of the Athenian fort there, and wrecked one of two ships that had been drawn up on the shore. At Peparethos likewise there was a certain recession of the waters, but no inundation; and there was an earthquake, which threw down a part of the wall as well as the prytaneum and a few other houses…"

(Thycydides,Lo.ii.157-9 (adapted))

### Diodorus Siculus, XII.59.1-2

"While the Athenians were busied with these matters, the Lacedaemonians, taking with them the Peloponnesians, pitched camp at the Isthmus with the intention of invading Attica again; but when great earthquakes took place, they were filled with superstitious fear and returned to their native lands. So severe in fact were the earthquakes in many parts of Greece that the sea actually over ran and destroyed some cities lying on the coast, while in Locris the isthmus of a peninsular which then existed was broken through and the island known as Atalante was formed."

(Diod.Sic. Lo.v.49 (adapted))

### Strabo I.3.20

"Demetrius of Callatis, in his account of all the earthquakes that have ever occurred throughout all of Greece, says that the greater part of the Lichades Islands [between Euboia and Lokris]and Cenaeum [opposite Lokris] was engulfed; the hot springs at Aedepsus and Thermopylae, after having ceased to flow for three days, began to flow afresh, and those at Aedepsus broke forth also at another source; at Oreus [in NE Euboia] the wall next to the sea and about seven hundred of the houses collapsed; and as for Echinus and Phalara and Heracleia in Trachis, not only was a considerable portion of them thrown down, but the settlement of Phalara was overturned, ground and all. And, says he, something quite similar happened to the people of Lamia and of Larissa; and Scarphia, also was flung up foundations and all, and no fewer than seventeen hundred human beings were engulfed, and over half as many Thronians; again a triple-headed wave rose up, one part of which was carried in the direction of Tarphe and Thronium, another part to Thermopylae, and the rest into the plain as far as Daphnus in Phocis; fountains of rivers were dried up for a number of days, and the Sphercheius changed its course and made the roadways navigable, and the Boagrius was carried down a different ravine, and also many sections of Alope, Cynus and Opus were seriously damaged, and Oeum, the castle above Opus, was laid in utter ruin, and a part of the wall of Elatia was broken down, and at Alponus, during the celebration of the Thesmophoria, twenty-five girls ran up into one of the towers at the harbour to get a better view, the tower fell, and they themselves fell with it into the sea. And they say, also, of the Atalanta near Euboea that its middle portions, because they had been rent asunder, got a ship-canal through the rent, and that some of the plains were overflowed even as far as twenty stadia, and that a trireme was lifted out of the docks and cast over the wall."

(Strabo, Lo.ii.223-7)

# Appendix 5: Practical Application of GIS in this Investigation

The practical aspects of the Geographical Information System (GIS) in terms of the basic GIS operations are outlined in this appendix. This includes an outline of the system, the software and the principle GIS operations (modules) used in the construction of the earthquake effects model and its application to the archaeoseismological methodology outlined in Chapter 3.

## 5.4 The System

GISs are a complex interaction between software and hardware; without operable hardware it is not possible to acquire the necessary data, and extremely difficult to develop a detailed high resolution database from which all further operations and applications originate. The software initially selected for use in this thesis was IDRISI for Windows Version 1 (hereafter abbreviated to IDRISI) supplied by Clark Laboratories, USA installed on a Pentium II 200 MHz laptop personal computer with 16Mb RAM and a 2Gb hard drive storage capacity (hardware). An Omega Zip and writeable CD drive were necessary for the storage of both raw data input files (specifically the satellite data) and output image files, several of which after manipulation and analysis exceeded the capacity of a standard 3.5" floppy disk drive. The digitising package (TOSCA) supplied with IDRISI was found to be inadequate for the quantity and capacity of digitising and encoding of the analog data layers required and alternative software was used.[70] However, non-indigenous (to the system) data acquisition created problems with data format incompatibility, which resulted in the need for complicated export procedures in the digitising software (for example ARCINFO, ERDAS and PCI WORKS[71]) and import into IDRISI. Subsequently, IDRISI due to its 8-bit data capacity, was found to be inadequate for the generation of the Digital Elevation Models that included negative values and real numbers. Therefore, DEMs requiring real numbers were generated in the ORTHOENGINE Version 6.2[72] software and imported into IDRISI. Unfortunately, the import procedure converts the real numbers into integers and thus reduces the ultimate vertical output resolution. However, the preference for creating data with real values means that future work using more sophisticated GIS software will retain both a high level of accuracy and resolution.

Irrespective of the brand of software selected, a GIS comprises the following four essential elements: (1) data acquisition and reprocessing; (2) data management including storage; (3) data manipulation and analysis; and (4) product generation or Output (Star & Estes, 1990). The first two elements (data acquisition and data management) are topic/project independent and are discussed in the remainder of this subheading. Although many of the operational modules for data analysis, manipulation and output are also project independent, they are included in following section 5.4.4 which discusses data manipulation in the context of this thesis research.

### 5.4.1 Data

Two data types are necessary to carry out an archaeoseismological appraisal; *spatial* and *temporal*. Spatial data refer to a location within space and are primarily geographical, whereas the temporal data are more usually attribute data, although in archaeology they are generally identified by their spatial location as well their as temporal position within a relative chronology. They are therefore primarily time related, although still maintaining a geographical reference. For example, a parcel of land can be identified spatially by the boundary co-ordinates and also by an attribute describing, for example, land use. The changes over time in land use can be analysed, and this represents the temporal element associated with the spatially defined and geographically referenced land parcel. Clearly, the division between spatial and temporal data in this context is less discrete and was specifically selected as an example for its analogy with archaeological data. However, it should be remembered that temporal information does not necessarily have to have a spatial element. Much historical or archaeological information focuses on the object and its development or decline; for example, the usage pattern of a settlement site can be discussed without reference to the geographical position, which is referred to as a historiographic interpretation in Chapter 2.

Within a GIS, spatial data is differentiated into one of the follow three categories: dimensionless *point* data, single dimension *line* data, and two dimensional *areas* (usually termed polygons). Due to the problems experienced with data acquisition (i.e. digitising), the majority of the acquired data levels, i.e. non-product information layers, in this investigation are either point or line data. It should be noted that the ultimate quality of the information products (output results) of any operations and transformations within a GIS is entirely dependent upon the quality (specifically the scale, accuracy and precision) of the acquired datasets used. Once encoded, datasets are termed *layers* and/or *coverages*. Therefore, only the highest quality of digitising accuracy and will result in high quality products.

---

[70] The author acknowledges the generosity of the Geography Department at Reading University, UK, for the use of the digitising facilities, and of Dr A. Ganas for technical assistance with previously unfamiliar hardware and software packages.

[71]71 These are the manufacturers names of alternative industry standard GIS and image processing software packages.

[72] ORTHOENGINE is a trademark of PCI GEOMATICS INC

| SHEET (SUBSET)[74] | SCALE (m) | PUB'D | EDITION DATE | DIGIT Z'D C/Y/P/S /N[75] | CONTOUR DEPTH INTERVAL (m)[76] | ERROR[73] X | Y | Z |
|---|---|---|---|---|---|---|---|---|
| KHALKIS | 1:250 000 | HAGS[77] | 1978 | C | 100 | ±75m | ±75m | ±25m |
| 1556 | N/K | BA[78] | 1890 | S | Variable (Fathoms[79]) | N/K | N/K | N/K |
| GREECE 1556 (East Coast) | ? | HO[80] | ? | Y | Variable | ? | ? | |
| LIVANATAI | 1:50 000 | HAGS | 1971 | Y | 20 | 15-25m | 15-25m | 5-10m |
| ATALANTI | 1:50 000 | IGME[81] | 1965 | ? | 20 | 15-25m | 15-25m | 5-10m |
| LIVANATAI (Opus) | 1:5 000 | HAGS[82] | 1977 | Y | 1-4 | 1.5-5m | 15-25m | 0.25-1m |
| LIVANATAI (Alai) | 1:5 000 | HAGS | 1977 | Y | 1-4 | 1.5-5m | 15-25m | 0.25-1m |
| LIVANATAI (Kynos) | 1:5 000 | HAGS | 1977 | ? | 1-4 | 1.5-5m | 15-25m | 0.25-1m |
| ATALANTI | 1:30 000 | ANG[83] | 1990 | ? | 20 | N/K | N/K | N/k |

**Table A5.1**: Cartographic data used in this GIS showing origin, scale, and errors. The errors given here are those specifically attached to the analog map format, and do not include the errors accumulated during encoding.

| DATASET | ORIGINATING FORMAT | IDRISI Yes/No |
|---|---|---|
| **Topography** | **Sheet Map** | **Yes** |
| **Drainage** | **Sheet Map** | **Yes** |
| **Roads** | **Sheet Map** | **Yes** |
| **Bathymetry** | **Sheet Map** | **Yes** |
| Tectonic Background | Text | No |
| Geology | Sheet Map/Fieldwork | No |
| Geotechnical Information | Sheet Map/Text | No |
| **Regional Seismicity (Contemp)** | **ASCII Text** | **Yes** |
| Historical Seismicity | Text | No |
| Ancient Seismicity | Text | No |
| Archaeological Seismicity | Text | No |
| Regional General Chronology | Text | No |
| Site Occupation Histories | Text | No |
| **Seismogenic Sources (Faults)** | **Sheet Map** | **Yes** |
| **Archaeological Sites** | **Sheet Map** | **Yes** |
| **Coastline** | **Sheet Map** | **Yes** |
| **Modern Settlements** | **Sheet Map** | **Yes** |

**Table A5.2**: A table of datasets collected within the empirical database. The first column describes the data set, the second gives the original format of the data and the third indicates whether the data set was included in the GIS. Note the spatial focus of incorporated data layers that reflects the application discussed.

---

[73] These are calculated from the published errors. X & Y are circular in the horizontal (i.e. planimetric). Z is linear in the vertical.

[74] Subset refers to a window of a larger scale sheet, arbitrarily named by the author according to the name of the archaeological site within the subset.

[75] C = Coast Only; Y = Topography or Bathymetry; P = Polygon data; S = Scanned for graphics only; N = Not digitised.

[76] Except where indicated in table.

[77] Hellenic Army Geographic Service.

[78] British Admiralty Maps supplied by United Kingdom Hydrographic Office, Taunton, England.

[79] 1 fathom = 6ft x 0.33 = 1.829m

[80] Hydrographic Office, Taunton, England.

[81] Institute of Geological and Mineral Exploration, Greece

[82] HAGS 1:5 000 are subsets reproduced from the relevant HMGS 1:50 000 data set by HAGS upon request.

[83] Angelides, 1992

*Scale, Accuracy & Precision*
Scale, accuracy and precision are often misinterpreted, and this subsection provides a general outline of each of the terms.

Scale is a ratio value used to provide meaning to diagrammatic representations of data. In terms of cartographic representations, it provides real earth distances for measurements on the sheet map. *Large scale* refers to maps that have a small distance ratio, i.e. 1:25,000 meaning that the map covers a small geographical area and provides high resolution data.[84] Conversely, *small scale* refers to maps that have a low resolution due to a greater geographical coverage, for example, a ground to map ratio of 1:100,000. As noted above, and in accordance with all computer applications, the quality of the output results mirrors the quality of the input data, and it is only possible to obtain high resolution products by using high resolution acquired data. Additionally, precision and accuracy need to be considered, especially when dealing with spatial data. Accuracy is the 'freedom from error or bias' (Star & Estes, 1990), whereas precision is the 'degree of accuracy of numerical representation' and in GIS principally refers to the number of digits following the decimal point (Burrough & McDonnell 1998).

### 5.4.2 Data Acquisition

The data sets collected in Chapter 4, including the metadata[85] originated in a variety of forms (e.g. graphic, textual, tabular) generally in non-encoded (i.e. non-digital) format (the LANDSAT image being the only exception), and required pre-processing prior to import into the GIS. This was also true for data which was being imported between different GIS software packages (i.e. from ARCINFO to IDRISI).

The original format of the data layers described in Chapter 4 is shown in Table A5.2, with the final column indicating those that were ultimately imported into IDRISI via the procedures outlined in the flow diagrams under the relevant subheadings. The predominance of cartographic data in the highlights the research objective, i.e. determining the validity of archaeological remains as proxy indicators.

*Digitising*
All the data layers used in this thesis were available only in non-encoded (or analog) format (excepting the satellite image) and as such required manual digitisation. Due to technical problems with the 'in house' digitising software

(TOSCA) and data format incompatibility between software packages, input of the data layers was restricted to those deemed essential for this research application and therefore have a predominantly topographic theme (see Table 5.2). Although different layers were digitised at different times, the procedure was essentially the same and is represented diagrammatically in Figure A5.1. Analog data (i.e. topography contours, hydrology, bathymetry, etc.) were manually digitised as vectors using LITES2 within the LASERSCAN software system, which were imported into ARCINFO V7.0.1 for co-ordinate transformations and finally exported into IDRISI using the ARC UNGENERATE (UNGEN) format.

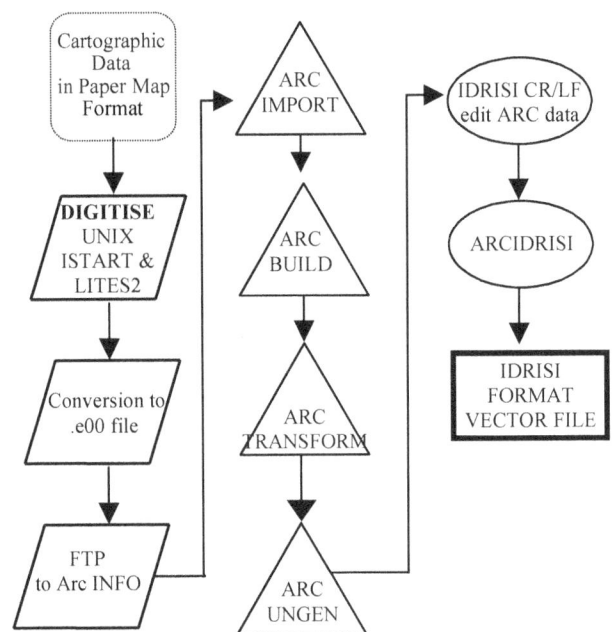

**FIGURE A5.1**: FLOW DIAGRAM SHOWING THE PROCEDURE FOR DATA INPUT FROM PAPER MAPS TO DIGITAL DATA IN IDRISI FORMAT. PARALLELOGRAMS INDICATE VMS SYSTEM MODULES. TRIANGLES REPRESENT ARCINFO UNIX MODULES AND IDRISI WINDOWS MODULES ARE INDICATED BY ELLIPSES. SOLID LINES INDICATE PROCEDURES WITHIN A SOFTWARE; DASHED LINES INDICATE THE USE OF INPUT/OUTPUT DATA SETS FOR DIFFERENT SOFTWARE.

*Digital Data Import*
Digital satellite images (LandSat 4) were used to provide an overview of the regional topography and also as a drape in the orthographic images of the Atalanti area to enhance the pseudo-three-dimensional view. Raw satellite data require specialised pre-processing prior to use, in terms of both product output (i.e. images) and importation into other geographical software packages. In this case the data were acquired in a pre-processed TIFF 8 bit format; therefore the method outlined in Figure A5.2 details only the procedure for import of satellite image files into IDRISI.

---

[84] Resolution is defined by Chrisman (1997, p 27) as the 'least detectable difference in a measurement', and applies to all components of the GIS: time, space and attribute.
[85] Data about data, such as comments about the dataset itself. For example, the name of the cartographer and date of drafting/publication usually printed on the sheet maps.

*Import of ASCII Text Data: Earthquake Data*[86]

The earthquake data used in this thesis were provided by the National Observatory in Athens in ASCII format, giving the date, time (in hours, minutes and seconds), location (in lat/long), depth of the focus, and magnitude of events within the specified field area. These data were edited in an ASCII text editor to remove events below the threshold of geological evidence, i.e. Ms 3.9 (McCalpin & Nelson 1996), and were then converted from the latitude longitude co-ordinates into geo-referenced UTM point vector data using the public domain software program PROJ 4.3 (specifying the international 1909 ellipsoid, zone 34 and the European 1950 datum). These data were then exported from ARCINFO in an UNGEN format and imported into IDRISI using the ARCIDRISI module.

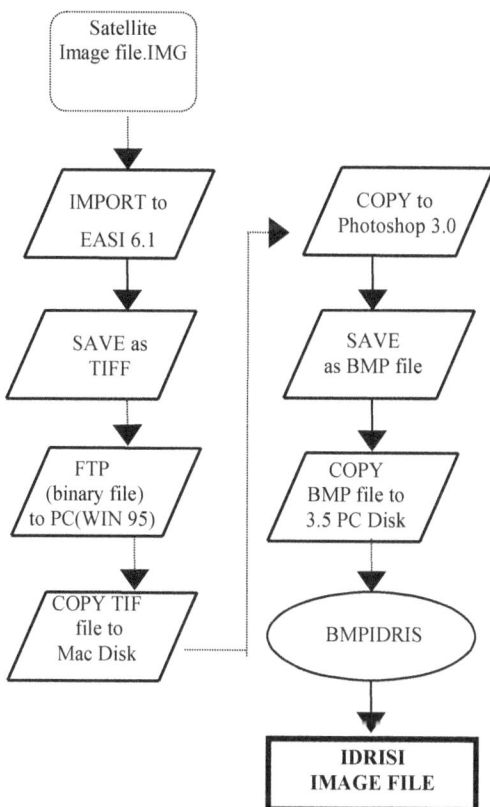

**FIGURE A5.2:** FLOW DIAGRAM OF THE PROCEDURE USED TO IMPORT PRE-PROCESSED DIGITAL SATELLITE DATA INTO IDRISI. THE ORIGINAL IMAGE FILE IS SHOWN IN ROUNDED SQUARE, NON IDRISI MODULES ARE SHOWN AS PARALLELOGRAMS, IDRISI MODULES AS ELLIPSES AND IDRISI OUTPUT FILE AS A HEAVY BOARDERED RECTANGLE. THE ORDER OF OPERATION IS GIVEN BY THE DASHED LINES WITH ARROWS

---

[86] The instrumental earthquake data was supplied by the National Observatory of Athens with the assistance of the director, Dr G Stavrakakis.

### 5.4 3 Manipulation of Data

As noted above (Section 5.2.1), a neutral theoretical model of the effects of an earthquake in an area can be derived by constructing a separate data layer for each element of the earthquake and then recombining them through the data manipulation modules available in the GIS. The construction of each element using the GIS is discussed in the following subsections.

*Deformation Ellipses (Zones of Influence (ZoI))*

As noted above (section 5.2.1), the amount of subsidence or uplift observed in an archaeological site can be quantified only where there is an independent horizontal datum, for example mean sea level as indicated by the coastline. Therefore, in terms of archaeoseismology, it is the spatial extent of deformation that is of primary importance as a guide to defining whether relative vertical movement in a site is the result of co-seismic slip or the product of other alternative mechanisms, such as coastal slumping or a eustatic change in sea level. Within this context, the mathematically derived deformation ellipse associated with co-seismic slip on a fault can be regarded as a Zone of Influence (ZoI) and as before is a geometrical configuration of the mathematical model (as discussed in section 4.6.1).

Using a GIS, zones of influence can be artificially created by operating the DISTANCE algorithm on a digitised source, i.e. a rasterised fault vector, and because deformation ellipses are topography independent (i.e. the movement up or down on a fault will occur irrespective of physiography) the spatial extent of the deformation associated with specific faults can be derived by overlaying a GIS derived zone of influence with a geo-referenced map, for example a rasterised topographic map or digital elevation model (DEM).

The flow diagram in Figure A5.3 shows the procedure used to derive the zones associated with the known seismogenic sources in the Gulf of Evvia. An output layer with distance rings increasing in value away from the source, was derived using previously digitised sources and the DISTANCE algorithm. The farthest extent of deformation associated with a single event on a specified fault is indicated at the point where there is no discernible change in ground elevation (Ma & Kusznir 1995).

Therefore, to generate the ZoI within the GIS, the values in the unlimited distance output image file (Figure A5.3) are reclassified using the RECLASS algorithm to show only two values: 1 for distances up to the point of zero vertical motion, and 0 for those outside. In this thesis the horizontal distance for the Atalanti fault was obtained from the deformation ellipse calculated by Ganas and Buck (1998), and zero vertical motion was noted at 20 km from the source.

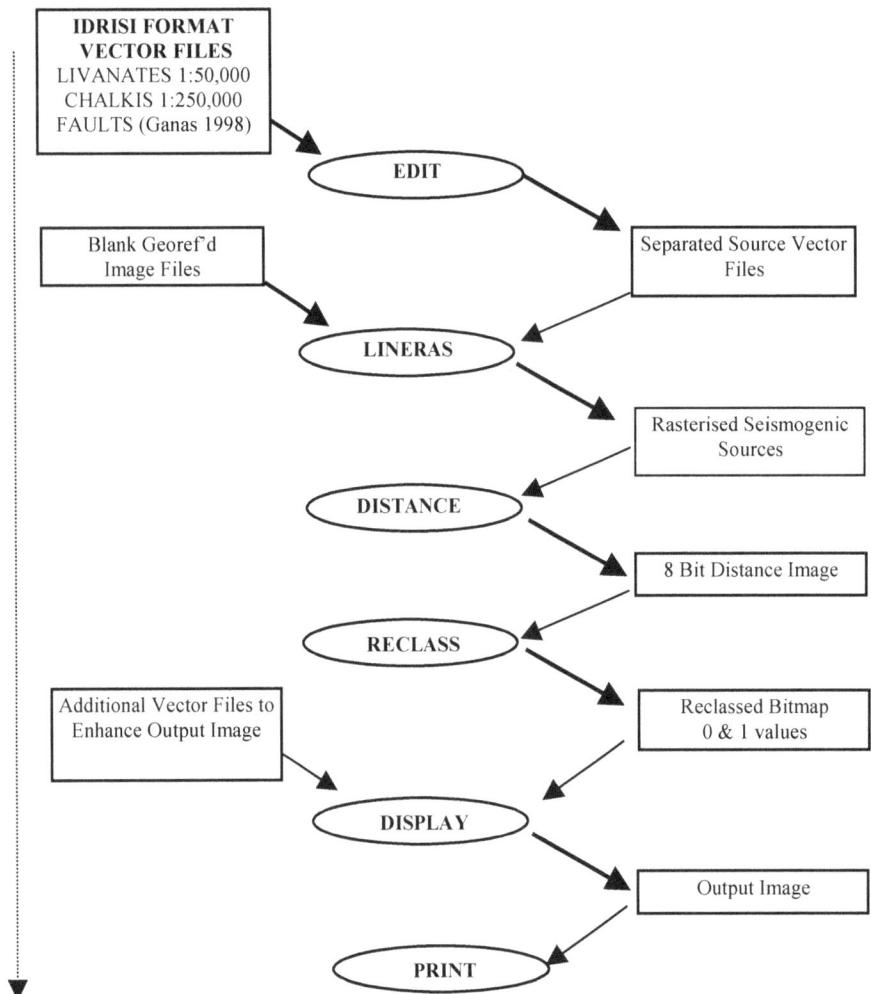

**FIGURE A5.3**: FLOW DIAGRAM ILLUSTRATING THE PROCEDURE FOR THE PRODUCTION OF THE ZONES OF INFLUENCE FROM THE PREVIOUSLY DIGITISED SEISMOGENIC SOURCES USING IDRISI. ELLIPSES INDICATE IDRISI OPERATION MODULES. RECTANGLES REPRESENT DATA FILES. THIN ARROWS INDICATE INPUT FILES AND THICK ARROWS INDICATE OUTPUT FILES. THE BROKEN ARROW ON THE LEFT INDICATES THE ORDER OF OPERATION. THE RESULTING OUTPUT FILE IS HIGHLIGHTED BY A THICK BORDER. NOTE THAT, AS WITH MANY MODULES IN IDRISI, THE OUTPUT FILE FROM THE PREVIOUS MODULE BECOMES THE INPUT OF THE SUBSEQUENT MODULE.

*Digital Elevation Models (DEM)*

Digital elevation models are images that store elevation (height) data on a grid and are frequently used to digitally create and visualise the topography of a study area. The flow diagram at Figure A5.4 shows the procedure used for the construction of the DEMs contained within this thesis. Two techniques were used to extract DEMs from previously digitised contours: (1) the IDRISI v1.01.002 technique, and (2) ORTHOENGINE v6.22. The IDRISI technique uses the INTERCON algorithm to interpolate the elevation contours that had previously been rasterised using the LINERAS module. The resulting faceted model was smoothed using a series of filter algorithms in the FILTER module. The most commonly used filter is a Mean (3 x 3) moving window. However, the extracted DEM contained many artefacts or errors (for example, splays, fans, depressions and elevation crosses) that were probably the result of the IDRISI interpolation error Therefore, an alternative software package was used to obtain a more accurate representation of the DEM. The UNGEN contour files and related attributes (e.g. heights)

were imported into ORTHOENGINE (OEAE)[87] and the program's interpolation algorithm was applied. The extracted DEM (Figure A5.12) contains no artefacts, and depicts well the morphological characteristics of the area, i.e. high topography and the low lying coastal area. The errors associated with the DEM extraction are given in table 5.3 and are the cumulative result of the following:

1. the individual map error (see Table 5.1);
2. the operators digitisation error (this is approximately 0.1mm X the map scale[88]);
3. co-ordinate transformation error (negligible, close to zero); and
4. the grid interpolation error (depending on the software and grid size).

---

[87] OrthoEngine Airphoto Edition.
[88] In the case of a 1:5,000 map the digitisation error is 0.1mm x 5m, or 50cm

Topographic visualisation of digital data can be enhanced through the use of a pseudo-perspective display technique that produces orthographic images, which can be constructed with any combinations of viewing azimuth and angle:[89] for example, looking from the north at a 45° viewing angle to the surface (see Figure A5.13 showing the Orthographic image of the Opus topography using these viewing parameters and with a satellite overlay).

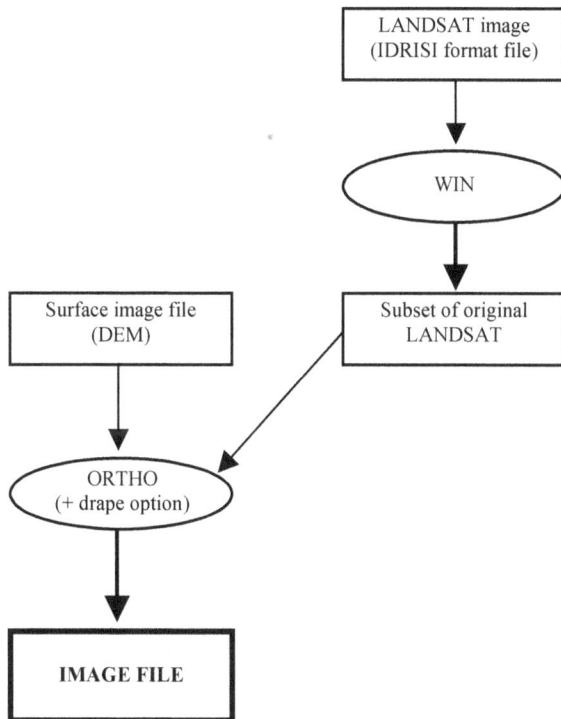

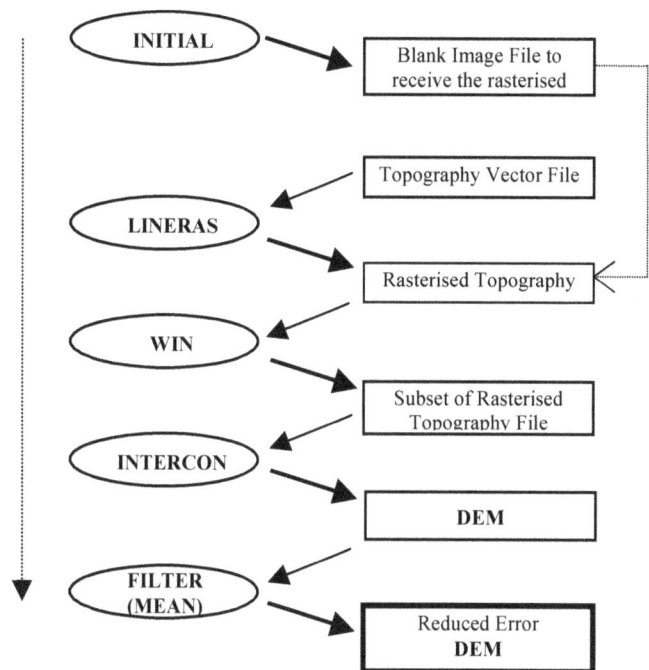

**FIGURE A5.4**: FLOW DIAGRAM ILLUSTRATING THE PROCEDURE FOR THE PRODUCTION OF A DIGITAL ELEVATION MODEL (DEM) USING IDRISI. ELLIPSES INDICATE IDRISI MODULES. RECTANGLES REPRESENT DATA FILES. THICK ARROWS INDICATE INPUT FILES AND THIN ARROWS INDICATE THE OUTPUT FILES. THE BROKEN ARROW ON THE LEFT INDICATES THE ORDER OF OPERATION. THE BROKEN ARROW ON THE RIGHT INDICATES THE POINT AT WHICH THE BLANK FILE CREATED IN INITIAL IS BROUGHT INTO THE PROCEDURE. THE RESULT FILE IS HIGHLIGHTED BY A THICK BORDER. NOTE THAT, AS WITH MANY MODULES IN IDRISI, THE OUTPUT FILE FROM THE PREVIOUS MODULE BECOMES THE INPUT OF THE SUBSEQUENT MODULE.

**FIGURE A5.5** FLOW DIAGRAM SHOWING THE PROCEDURE FOR PRODUCING 3D ORTHO IMAGES WITH SATELLITE DRAPES. IDRISI MODULES ARE REPRESENTED BY ELLIPSES, INPUT FILES ARE IN BOXES WITH AN ARROW, OUTPUT FILES INDICATED BY A BOLD ARROW. SATELLITE IMAGE IS LANDSAT 4, 28 JANUARY 1988 PRE-PROCESSED BY A. GANAS (1998) USING THE ENHANCED CHANNEL 5 (SHORTWAVE INFRA-RED) IMPORTED DIRECTLY INTO IDRISI AS AN IMAGE FILE.

*Orthographic images*
Orthographic images (see Figure, A5.13) were created through the operation of the ORTHO module using the extracted DEMs as the input image. The visual effect of topography was created by draping a satellite image and using a specific palette that showed high topography in purple and the low lying shore areas in yellow/green.

The easy production of the orthographic images from derived DEMs, i.e. the simple operation of a pre-existing IDRISI module, ORTHO, means that new images can be created to view a study area from a variety of perspectives, enhancing the understanding of an archaeological site within the topographical setting. This can be carried further by adding other data layers, such as rivers, roads and archaeological sites, to display such aspects as drainage, communications and sites within the environment.

| DEM | Planimetric X & Y | Vertical (Z) |
|---|---|---|
| 1:5,000 Opus & Alai | 1 – 1.5m | 25 – 30cm |
| 1:50,000 Livanates | 15 – 20m | 4 – 5 m |

**Table A5.3:** Total error associated with the DEMs extracted using ORTHOENGINE

---

[89] Viewing angle refers to the inclination of the eye to the image plane.

*Slope Maps*

Slope maps are a useful data layer to be considered in assessing the stability of the ground upon which sites are located, and are therefore directly associated with building an overview of the seismic hazard of an area or locality. Previously, slopes have been manually calculated and maps constructed from the topographic diagrams or fieldwork measurements (e.g. Angelides, 1990). However, within the GIS such maps can be simply and accurately constructed from a previously extracted DEM using the standard SLOPE module within the GIS. The algorithm calculates elevation change versus distance in relation to the neighbouring pixels in either percentages or degrees (operator specified). In this thesis, slope maps were initially produced from the IDRISI DEMs using the slope option within the IDRISI SURFACE analysis module. However, the resulting output output maps contained a number of errors due to the presence of artefacts (as noted in the previous section) within the input DEM. Therefore, the slope maps used for analysis in this thesis were derived using the SLOPE algorithm on the DEMs created within in the ORTHOENGINE software and imported via the TIFIDRIS operation.

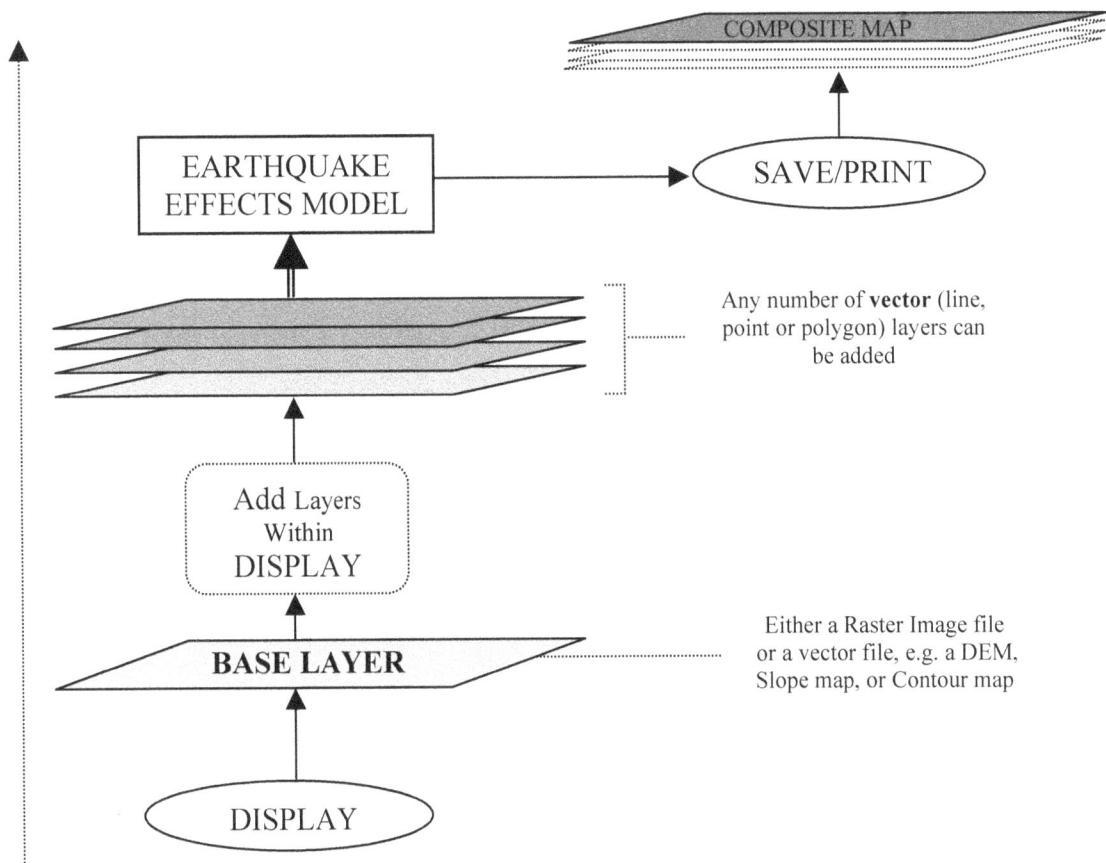

**FIGURE A5.6:** DIAGRAM SHOWING THE PROCEDURE FOR THE CONSTRUCTION OF THE EARTHQUAKE EFFECTS MODEL. IDRISI MODULES ARE INDICATED BY AN ELLIPSE, IDRISI FILES BY A RECTANGULAR BOX AND OPERATIONS WITHIN A MODULE BY A DASHED BOX. THE 'EARTHQUAKE EFFECTS' MODEL IS DERIVED BY ADDING ANY NUMBER OF LAYERS WITHIN THE DISPLAY MODULE. THIS COMPOSITE IMAGE CAN THEN BE EITHER SAVED (AS A BMP FILE, SCREEN DUMP FOR EXPORT, OR NEW IDRISI IMAGE FILE) FOR FUTURE USE OR PRINTED DIRECTLY.

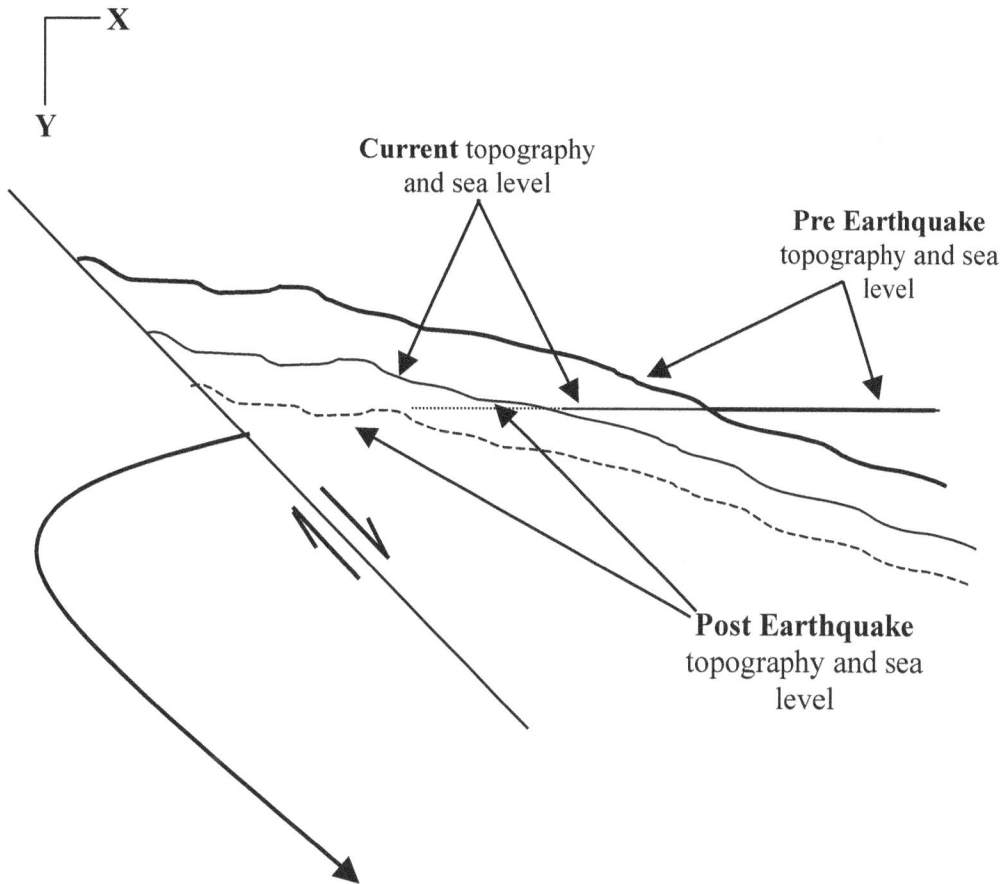

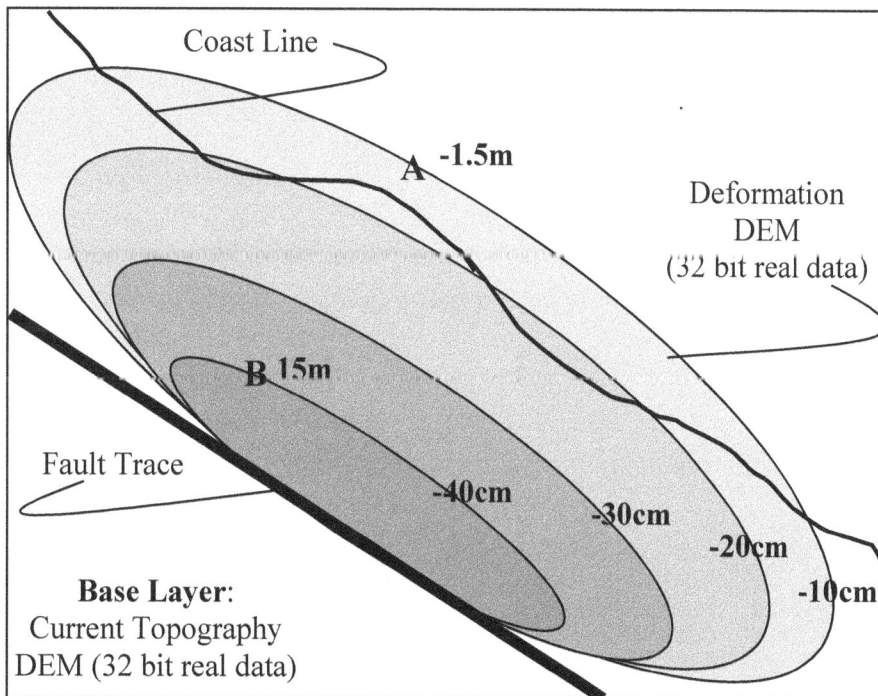

**FIGURE A5.7**: **A**: SCHEMATIC REPRESENTATION (IN CROSS-SECTION) OF THE RELATIONSHIP OF PRE- AND POST-EARTHQUAKE TOPOGRAPHY TO THE PRESENT DAY TOPOGRAPHY. **B:** A PLAN VIEW REPRESENTATION OF RELATIONSHIP BETWEEN THE BASE DEM AND THE DEFORMATION ELLIPSE DEM (BOTH IN 32 BIT REAL DATA). THE ALGORITHM TO PRODUCE THE PRE AND POST EARTHQUAKE DEMs IS GIVEN IN FIGURE A5.8.

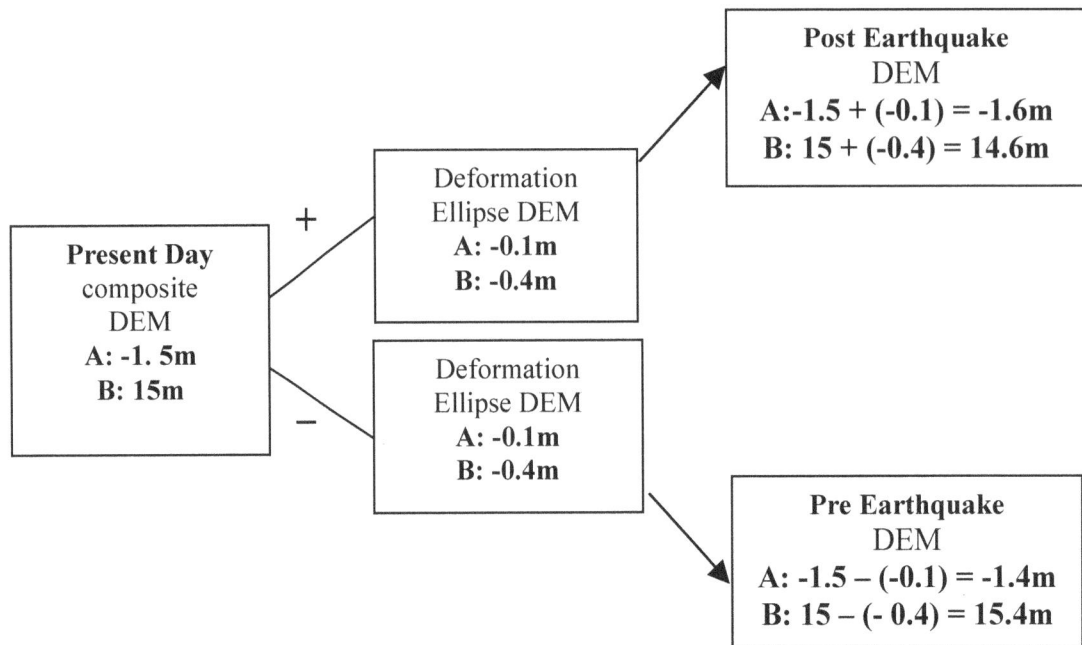

**FIGURE A5.8:** THE PROCEDURE SUGGESTED IN THIS THESIS TO PRODUCE PRE- AND POST-EARTHQUAKE DEMs. FOR BOTH RESULTING DEMs THE GIS METHOD EFFECTIVELY REBUILDS OR PREDICTS THE TOPOGRAPHY BY ADDING TWO 32 BIT REAL DATA DEMs WITHIN THE OVERLAY MODULE. IT WAS NOT POSSIBLE TO EXTRACT THE NEW DEMs BECAUSE IDRISI (V.1) WILL NOT IMPORT 32 REAL (OR 16-BIT SIGNED) DATA. THEREFORE, THIS PROCEDURE HAS BEEN ILLUSTRATED IN THE DIAGRAMS IN FIGURE 5.19, AND NUMERICALLY DEMONSTRATED USING ARBITRARY ELEVATIONS IN THE EXAMPLES A AND B ABOVE.

### 5.4.5 Product Output

Once primary datasets have been imported into a GIS system, information products can be output on a user defined basis using data layers, either primary digital datasets or the derived output layers from data manipulation modules. For the purposes of this thesis, the primary use of the GIS was to construct a theoretical model of the effects of earthquakes at archaeological sites within the study area which would assist in making a qualitative assessment of the value of archaeological sites as indicators of past seismic activity.

*Constructing the Model of Earthquake Effects*
The model of the effects of earthquakes was constructed using a number of data layers (both raster and vector) to produce a single output product, using the procedure outlined in Figure A5.6. Any number of vector files, for example the geotechnical information (polygon vector file), site location (point vector file), seismogenic source vector lines and true deformation ellipse contours (line vector file), can be added to either a raster base layer, i.e. an extracted DEM, or a vector base layer, for example the digitised contours. This allows for the integration and visualisation of several datasets as coverages in order to facilitate an archaeoseismological assessment of a particular site in that they allow the effects of large magnitude earthquakes on archaeological sites, to be mapped in the Atalanti Plain.

*Pre- and Post- Earthquake Digital Elevation Models*
Pre- and post-seismic digital elevation models can be created by adding or subtracting the deformation DEM (a raster image of a deformation map, obtained either as a direct import of raster data or by interpolating digitised contour vectors within IDRISI) within the OVERLAY operation in IDRISI. For areas that contain only topography (and therefore only positive values), see example B in the procedure flow diagram shown in Figure. A5.8, that is also schematically represented in Figure A5.7).

Where a study area contains a coastline and sea (as is the case in Atalanti), it is first necessary to obtain a composite area DEM that includes both positive values (topography) and negative values (bathymetry). The composite DEM of the Atalanti Plain was produced in ORTHOENGINE as 32 bit real data. However, this was impossible to import into IDRISI due to incompatibility of data types (i.e. IDRISI v.1 only accepts 8-bit data). An overview is presented at Figure A5.15. The image was created by combining two separate contour vector files (one for topography and one for bathymetry) within the ORTHOENGINE software. This vector file was converted to a raster image and interpolated.

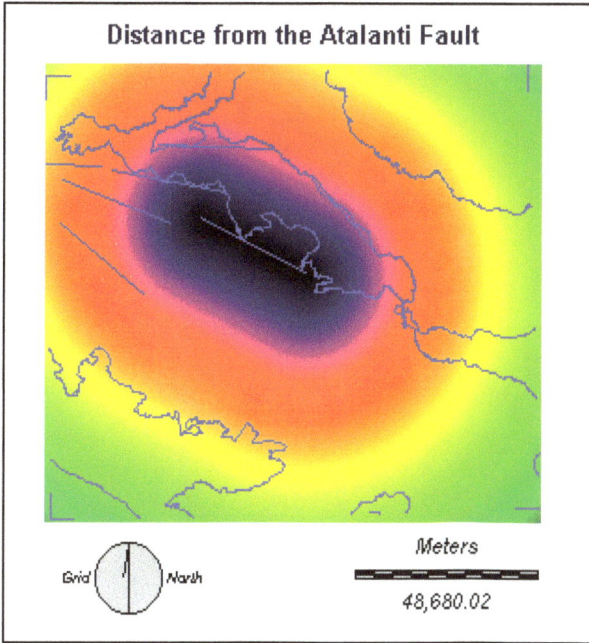

**Distance from the Atalanti Fault**

Figure A5.9: Output image showing the distance algorithm on the Atalanti Fault with a vector overlay of the coast and known seismogenic sources on the southern margin of the Gulf of Evvia.

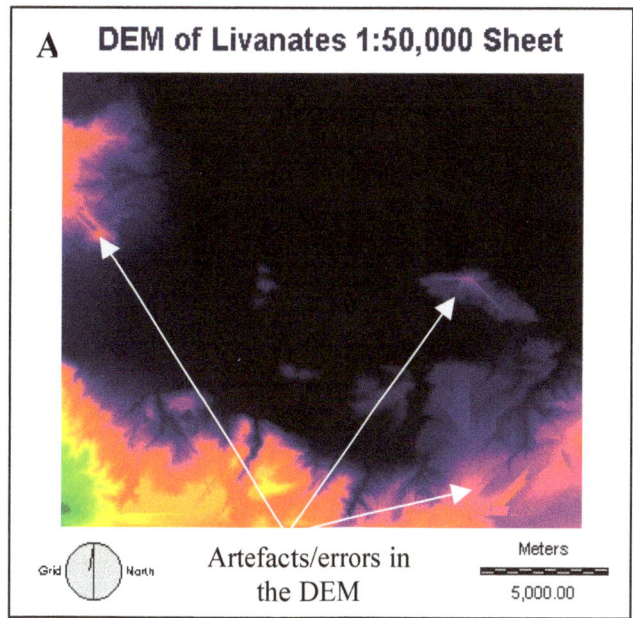

**Distance Image Reclassed to show 20km Zone**

**Figure A5.10**: Above image RECLASSed to define a 20 km zone of influence with an overlay of the coast and faults in blue. Pixels inside the zone are reassigned a value of 1 (white), those outside have a value of zero (black).

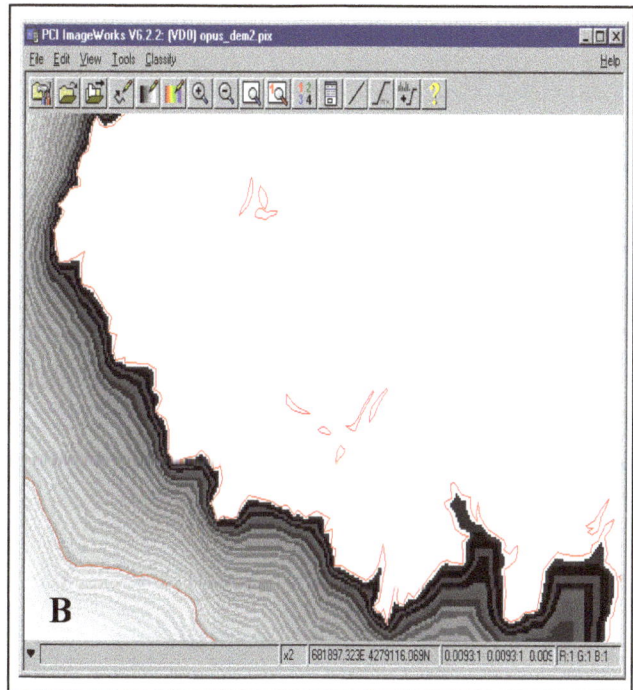

**A  DEM of Livanates 1:50,000 Sheet**

Artefacts/errors in the DEM

**Figure A5.11A**: DEMs of the Livanates 1:50,000 sheet. Note that the artefacts (errors) visible in IDRISI image are absent in the OEAE image extract (B) below.

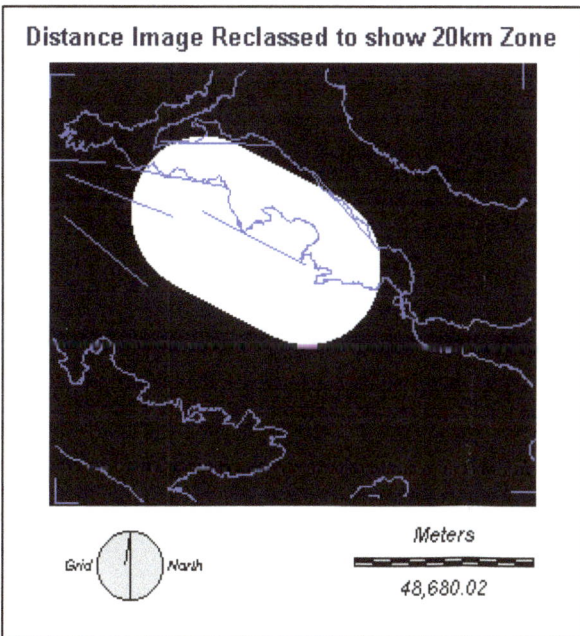

**Figure A5.11**: OEAE extracted DEM of the Livanates 1:50,000 sheet. Note that the artefacts (errors) visible in IDRISI image (A5.11A) above are absent this image

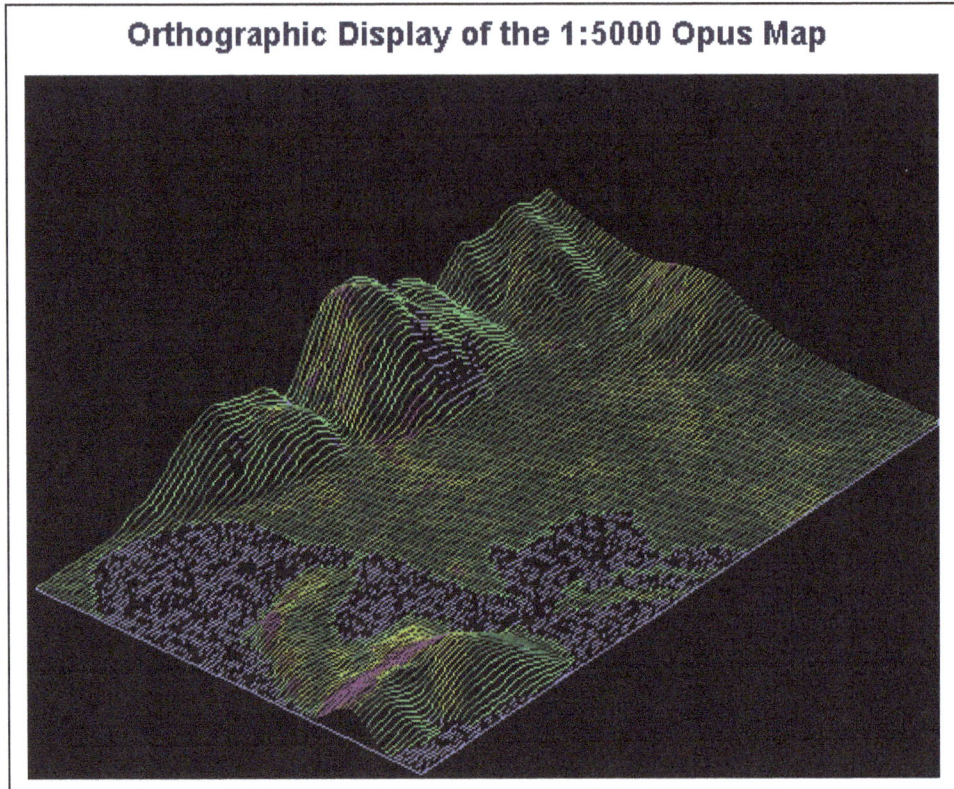

**Figure A5.13**: Orthographic display with a operator specified palette of the 1:5,000 Almyra (Opus) map. Viewing direction from the north-east on a 45o inclination. The sharp contrast of the range front (Khlomon mountains) and the Atalanti plain can clearly be seen, with Donkey Island in the foreground and the sea shown in blue. The colours are a function of the satellite drape image (i.e. reflectance values) rather than true attribute data.

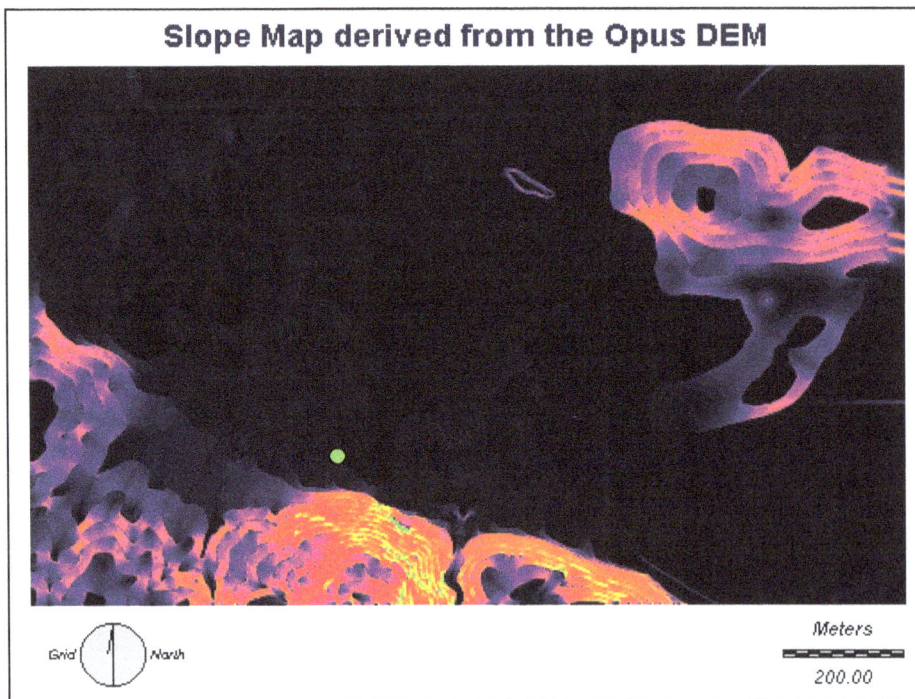

**Figure A5.14**: 8 Bit slope map derived from the Opus DEM (3m resolution). Values in degrees range from 0 (black), 10 – 15 (purples), 16 – 25 (pinks/reds) and 30 – 50 (oranges/yellows/greens). The green dot marks the site of Kyparissi (settlement) where the deformed foundations are located.

# Appendix 6: Peak Ground Acceleration and Intensity Calculations

## A6.1: R Value Calculations

The R value is the shortest distance from the site to the energy source (hypocentre) and is calculated using the following equation by Cornell (1968) in Ganas (1997). The values of $\Delta$ (where $\Delta$ is the distance from the site to the epicentre) and $h$ (the depth to hypocentre) were derived using trigonometry and can be seen in the diagram below (figure, A6.1 and A6.2).

$$R = \sqrt{\Delta^2 + h^2}$$

To find $\Delta$ and $h$:

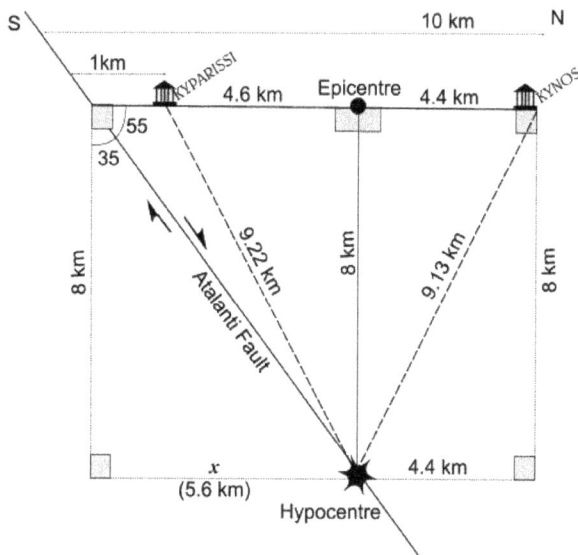

**Figure A6.1:** The geometrical associations of the earthquake and sites of Kyparissi and Kynos.

**To find $x$:**

$$\tan 35 = x/8$$
$$\therefore x = 8 \tan 35$$
$$= 8 \times 0.7$$
$$= 5.6 \text{ km}$$

*Kynos:*

$$R = \sqrt{8^2 + (10 - 5.6)^2}$$
$$\sqrt{64 + 19.36}$$
$$9.13 \text{ km}$$

*Kyparissi:*

$$R = \sqrt{8^2 + (5.6 - 1)^2}$$
$$\sqrt{64 + 21.16}$$
$$9.22 \text{ km}$$

*Alai:*

Assuming the epicenter is located at Skala (see calculations above), the closest distance to the fault hypocentre can be calculated by measuring the distance on the map from the Skala to the site of Allai (11km) as shown in the diagram at Figure A6.2.

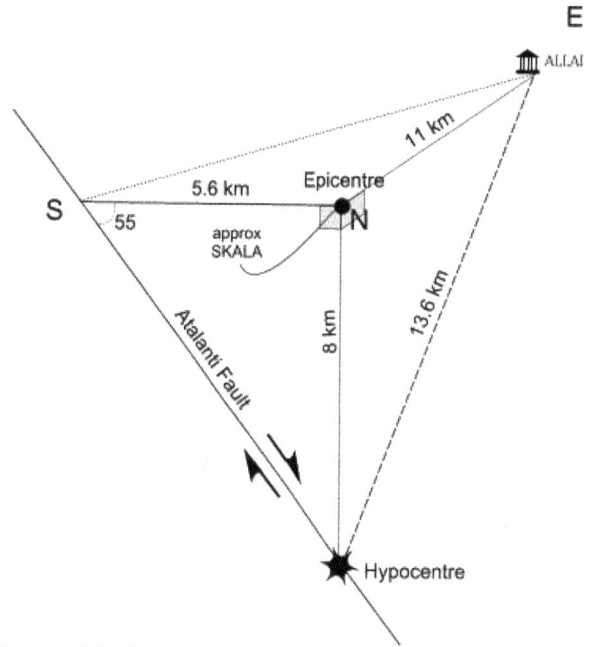

**Figure A6.2:** Geometrical associations of the hypocentre and Alai.

$$R = \sqrt{8^2 + 11^2}$$
$$\sqrt{64 + 121}$$
$$13.6 \text{ km}$$

## A6.2: Intensity

Earthquake intensity, **I**, was calculated for each site using the 1984 relational equation for shallow earthquakes in Greece by Papaioannou in Papazachos & Papazachou (1997), where **R** is distance in km from the hypocentre, and $M$ is the magnitude. The calculations assume an earthquake of similar magnitude of the 1894 Atalanti earthquake sequence, **M= 7**

$$I = 6.59 + 1.18\, M - 4.5 \log ( R + 17 )$$

*Kynos:*

$$I = 6.59 + 1.18 \times 7 - 4.5 \log ( 9.13 + 17 )$$
$$(6.59 + 8.26) - 6.37$$
$$8.48 \text{ or } \textbf{VIII} - \textbf{IX}$$

*Kyparissi:*

$$I = 6.59 + 1.18 \times 7 - 4.5 \log ( 9.2 + 17 )$$
$$(6.59 + 8.26) - 6.38$$
$$8.47 \text{ or } \textbf{VIII} - \textbf{IX}$$

*Alai:*

$$I = 6.59 + 1.18 \times 7 - 4.5 \log ( 13.6 + 17 )$$
$$(6.59 + 8.26) - 6.68$$
$$8.17 \text{ or } \textbf{VIII} - \textbf{IX}$$

## A6.3 Peak Ground Acceleration

### *Theodulidis (1991)*

Peak horizontal ground acceleration was calculated using the equation by Theodulidis (1991) in Papazachos & Papazachou 1997, p 124. R is the shortest distance from the site and the hypocentre and M= 7.

$$\text{Log } \alpha = 1.77 + 0.49\,M - 1.65 \log (R + 15)$$

### *Makropoulos (1985)*

A second peak ground acceleration was calculated using the equation by Makropoulos (1985). **R** is the shortest distance from the site and the hypocentre and **M**= 7.

$$\text{Ln } \alpha = 7.68 + 0.70\,M - 1.80 \ln (R + 20)$$

*Kynos:*

$$\text{Log } \alpha = 1.77 + (0.49 \times 7) - 1.65 \log (9.13 + 15)$$
$$= 1.77 + 3.43 - 2.28$$
$$= 2.918$$
$$= 10^{2.918}$$
$$\alpha = \textbf{827.94 gals}$$

*Kynos:*

$$\text{Ln } \alpha = 7.68 + (0.70 \times 7) - 1.80 \ln (9.13 + 20)$$
$$= 7.68 + 4.9 - 6.09$$
$$= 6.49$$
$$= e^{6.49}$$
$$\alpha = \textbf{658 gals}$$

*Kyparissi:*

$$\text{Log } \alpha = 1.77 + (0.49 \times 7) - 1.65 \log (9.2 + 15)$$
$$= 1.77 + 3.43 - 2.283$$
$$= 2.917$$
$$= 10^{2.917}$$
$$\alpha = \textbf{826.03 gals}$$

*Kyparissi:*

$$\text{Ln } \alpha = 7.68 + (0.70 \times 7) - 1.80 \ln (9.2 + 20)$$
$$= 7.68 + 4.9 - 6.07$$
$$= 6.51$$
$$= e^{6.51}$$
$$\alpha = \textbf{672 gals}$$

*Alai:*

$$\text{Log } \alpha = 1.77 + (0.49 \times 7) - 1.65 \log (13.6 + 15)$$
$$= 1.77 + 3.43 - 2.40$$
$$= 2.869$$
$$= 10^{2.869}$$
$$\alpha = \textbf{739.60 gals}$$

*Alai:*

$$\text{Ln } \alpha = 7.68 + (0.70 \times 7) - 1.80 \ln (13.6 + 20)$$
$$= 7.68 + 4.9 - 6.32$$
$$= 6.26$$
$$= e^{6.26}$$
$$\alpha = \textbf{523 gals}$$

**Note** the differences in the ground acceleration values (almost a magnitude) is due to the different equation parameters.

# Appendix 7: Quantifying Earthquakes

## Magnitude

The most common measurement of magnitude is the Richter Scale, is also known as the local magnitude ($M_L$), is the logarithm to the base 10 of the maximum seismic wave amplitude, in thousands of a millimeter, recorded on a (Wood Anderson) seismograph at a distance of 100km from the earthquake epicentre (Yeats *et al.* 1997). Therefore, in order to calculate the Richter Magnitude it is necessary to know both the distance and the amplitude parameters of the earthquake and as this quantification is not tied to any fundamental physical earthquake parameters, it essentially represents the size of an earthquake (*at its source*) rather than energy release. The scale was initially designed so that the smallest earthquakes would have a positive magnitude (hence a zero level earthquake was specified as the base level). However, new more sensitive instruments have since been developed and employed, and it is now possible to generate a negative event, i.e. a person with a sledge hammer in close proximity to a sensor (Brumbaugh, 1999).

## Intensity Scales.

Intensity scales measure the violence of earthquake shaking at a specified point based on the descriptions of earthquake phenomena, or damage. For this reason intensity scales are commonly described as measuring the feeling of the seismic event and as such are highly subjective as the description of the event is based upon the recognition of observable effects rather than any precise quantitative measurements. This coupled with the descriptive nature of intensity scales means that there are no fractional steps and intensity is labelled only by whole numbers. A number of intensity scales have been proposed (two are given below) and are used throughout the world. However, perhaps the most common used, and certainly the one used in Greece, is the Modified Mercelli Scale (Table A7.1).

### *Rossi-Forel Scale of Earthquake Intensity*

I. Recorded by a single seismogram or by some seismographs of the same model but not by several seismographs of different kinds; the shock felt by an experienced observer

II. Recorded by seismographs of different kinds; felt by a small number of people at rest

III. Felt by persons at rest; strong enough for the duration or the direction of shaking to be appreciable.

IV. Felt by persons in motion; disturbance of movable objects, door, windows; cracking of ceilings.

V. Felt generally by everyone; disturbance of furniture and beds; ringing of some bells.

VI. General awakening of those asleep; general ringing of bells, oscillation of chandeleirs, stopping of clocks; visible diturbance of trees and shrubs. Some startled persons leave their dwellings.

VII. Overthrow of moveable objects, fall of plaster, ringing of church bells, general panic, without damage to buildings.

VIII. Fall of chimneys, cracks in walls of buildings.

IX. Partial or total destruction of some buildings.

X. Great disaster, ruins, disturbance of strata, fissures in the earth's crust, rock falls from mountains

### *The Modified Mercelli Intensity Scale of Ground Shaking*

I. Not felt except by a very few under especially favourable circumstances

II. Felt only by a few persons at rest, especially on upper floors of buildings. Delicately suspended objects may swing.

III. Felt quite noticeably indoors, especially on upper floors of buildings, but many people do not recognise it as an earthquake. Standing motor cars may rock slightly. Vibration like a passing truck. Duration estimated.

IV. During the day felt indoors by many, outdoors by few. At night some awakened. Dishes, windows, doors disturbed; walls made cracking sound. Sensation like heavy truck striking building. Standing motor cars rocked noticeably.

V. Felt by nearly everyone; many awakened. Some dishes, windows, etc., broken; a few instances of cracked plaster; unstable objects overturned. Disturbance of trees, poles, and other tall objects sometimes noticed. Pendulum clocks may stop.

VI. Felt by all; many frightened and run outdoors. Some heavy furniture moved; a few instances of fallen plaster or damaged chimneys. Damage slight.

VII. Everybody runs outdoors. Damage negligible in buildings of good design and construction; slight to moderate in well-built ordinary structures; considerable in poorly built or badly designed structures; some chimneys broken. Noticed by persons driving motor cars.

VIII. Damage slight in specially designed structures; considerable in ordinary substantial buildings with partial collapse; great in poorly built structures. Panel walls thrown out of frame structures. Fall of chimneys, factor stacks, columns, monuments, walls. Heavy furniture overturned. Sand and mud ejected in small amounts. Changes in well water. Disturbs persons driving motor cars.

IX. Damage considerable in specially designed structures; well designed frame structures thrown out of plumb; great in substantial buildings, with partial collapse. Buildings shifted off foundations. Ground cracked conspicuously. Underground pipes broken.

X. Some well built wooden structures destroyed; most masonry and frame structures destroyed with

foundations; ground badly cracked. Rails bent. Landslides considerable from river banks and steep slopes. Shifted sand and mud. Water splashed (slopped) over banks.

XI. Few (if any) masonry structures remain standing. Bridges destroyed. Broad fissures in the ground. Underground pipes completely out of service. Earth slumps and land slips in soft ground. Rails bent greatly.

XII. Damage total. Waves seen on the ground surfaces. Lines of sight and level distorted. Objects thrown upwards into the air.

After (Brumbaugh, 1999)

| Richter Magnitude | Max Modified Mercelli Intensity | Typical Effects |
|---|---|---|
| 2.0 and under | I -II | Not generally felt by people |
| 3.0 | III | Felt by some indoors, no damage |
| 4.0 | IV - V | Felt by most, objects disturbed, no structural damage |
| 5.0 | VI - VII | Some structural damage, i.e. cracked walls and chimneys |
| 6.0 | VII - VIII | Moderate damage, i.e. fractures in weak walls and topples chimneys |
| 7.0 | IX - X | Major damage, i.e. collapse of weak buildings and cracking in strong buildings |
| 8.0 and over | XI - XII | Damage total or almost total |

**Table A7.1**: Comparative table of Richter and Modified Mercelli scales, with a short listing of typical effects (After Yeats *et al.*,1997, p 73).

# Appendix 8: Historical Background

## General Background the Settlement Patterns in Opuntian Lokris

### Before Christ Dates (B.C)

A note on dates: Many of the time brackets given are derived from generalised chronological histories, and for the earlier periods the actual dates will vary from site to site. The overall chronology is derived from an amalgamation of several source books, which for Greece as a whole are (Demand, 1996; Martin, 1996; Morkot, 1996). The historical background for Lokris is derived from (Fossey, 1990) with additional references cited in the text where appropriate.

| | |
|---|---|
| **18,000 - c15,000**<br>*Opuntian Lokris:* | **Palaeolithic**<br>To date no published details on Palaeolithic sites have been located. However, this does not confirm the absence of such sites in the field area, and it may be that sites do exist but as yet are undiscovered or not fully published in an accessible format. |
| **7,000 - 3,000**<br>*Opuntian Lokris:* | **Neolithic**<br>Fossey (1992) notes an overall slow increase in settlements during this period. Three sites are known in the Neolithic, Alai, Bazaraki, and Mitrou, and Fossey suggests that these sites are an extension of the Neolithic community in the north east bay of the Kopiac Lake (now Boiotia) and are sites such that they allow for ready access to the sea and a primary Neolithic food source, fish. |
| **3,200 - 1,000**<br>*Opuntian Lokris:* | **Bronze Age**<br>Overall settlement stability is demonstrated during the transition from Early to Middle Bronze Age in Lokris (Fossey 1992). However, a notable increase (almost double) in settlement is noted for the Middle to Late Bronze Age. Again the known sites occupy mostly coastal localities until the Late Bronze Age, when there is a switch to low hill sites, and Fossey (1992) notes that there is no loss of sites. The lack of inland sites may indeed be a function of preservation, as the fertile plains of the area are extensively cultivated. An apparent hiatus between the Middle and Late Helladic has been interpreted as a late inclusion of the area into the Mycenean world, and has been described as a 'culture lag' (Fossey 1992, p 104). |
| (Postpalatial)<br>**@1,200 - 750**<br>*Opuntian Lokris* | **The Dark Ages (Early Iron Age)**<br>There appear to be no evidence of settlement in Lokris during the Submycenean, scant indication of the Protogeometric, but a significant rise in the sites (or re-occupation of existing sites) during the remaining Geometric Period, with the beginning of settlement inland, specifically the corners of the Atalanti plain. |
| **c. 750 - 500**<br>*Opuntian Lokis* | **The Archaic**<br>During the Archaic the number of settlement expanded to a peak exceeding that of the Late Helladic with colonial activity, specifically in the establishment of Lokroi Epizephyroi in Sicily (Fossey 1992, p 108). |
| **500 - 323**<br>*Opuntian Lokris* | **Classical Age**<br>Population expansion is attested in this region by the slow and gradual continuation of settlement increase, with occupation of both the lower and upland slopes of Mnt Khlomon (i.e. Kastri Kolakas and Kastri Atalantis) Fossey (1992, p 113) notes that 'although Opuntian Lokris took part in Xerxes' invasion (Persian Wars) until the disaster at Thermopylai she obviously capitulated thereafter and was saved the depredations which the Persians visited upon neighbouring Phokis' which he interprets as the area having been spared any potential impediment to settlement development. Fossey (1992, p 113) also notes that the 426 earthquake did not seem 'to have effected settlement pattern or density' |
| **323 - 30**<br>*Opuntian Lokris* | **The Hellenistic period.**<br>There is a continuation of settlement from the Classical into the Hellenistic with preference for sites with harbour or low, eminent positions inland, with a plentiful supply of fresh water. Fossey (1992) notes that defence does not seem to have been a major factor, although several sites do take advantage of natural acropolis sites (for example Kyparissi and Kynos). |
| **146 - AD 330**<br>*Opuntian Lokris* | **The Roman Period**<br>There is slight evidence of this period in the archaeology; however, there is a slow decline in the number of settlements, particularly the sites that had expanded into the uplands. Similar inhabitation in the Imperial and Late Roman implies continuity of population |
| **AD 600 - AD 1100** | **Byzantine Greece** |
| **AD 1460 - AD 1831** | **Dark Ages** |

# References

1894a. Earthquakes in Greece. In: *Levant Herald and Eastern Express*, London.

1894b. Severe Earthquakes in Greece. In: *The Times*, London.

1918. *The Mediterranean Pilot*. His Majesty's Stationary Office, London.

Allen, K. M. S. 1990. Approaching Archaeological Space: an Introduction to the Volume. In: *Interpreting Space: GIS and Archaeology* (edited by Allen, K. M. S., Green, S. W. & Zubrow, E. B. W.). Taylor Frances, London.

Ambraseys, N. 1971. Value of historical records of earthquakes. *Nature* 232(August 6), 375-379.

Ambraseys, N. 1972. Earthquake hazards and emergency planning. *Build Internationa l*(January/February 1972), 38-42.

Ambraseys, N. 1973. Earth sciences in archaeology and history. *Antiquity*, 229-231.

Ambraseys, N. 1983. Notes on historical seismicity. *Bulletin of the Seismological Society of America* 73(6), 1971-1920.

Ambraseys. N. 1996. Seismicity of Central Greece. In: *Archaeoseismology* (edited by Stiros, S. & Jones, R. E.) Fitch Laboratory Occasional Paper Number 7. British School at Athens, Athens, 23-36.

Ambraseys, N. N. & Finkel, C. F. 1990. The Marmara Sea earthquake of 1509. *Terra Nova*, 167-173.

Ambraseys, N. & Jackson, J. A. 1990. Seismicity and associated strain of central Greece between 1890 and 1988. *Geophysical Journal International* 11, 663-703.

Ambraseys, N. & Karcz, I. The Earthquake of 1546 in the Holy Land. *Terra Nova* 4, 253-262.

Ambraseys, N. N. & White, D. P. 1996. Seismicity of the East Mediterranean and Middle East - I. Imperial College, London.

Amiran, D. H. K., Arieh, E. & Turcotte, T. 1994. Earthquakes in Israel and Adjacent Areas: Macroseismic Observations since 100 BCE *Israel Exploration Journal* 44 (3-4), 260 - 305.

Angelides, C. 1990. Geotechnical Report of the Atalanti Plain. Unpublished Doctoral thesis.

Armijo, R., Meyer, B., King, G. C. P., Rigo, A. & Papanastassiou, D. 1996. Quaternary evolution of the Corinth Rift and its implications for the Late Cenozoic evolution of the Aegean. *Geophysical Journal International* 126, 11-53.

Bacchielli, I. 1995. A Cyrenaica earthquake post 364 AD: written sources and archaeological evidences. *Annali di Geofisica* XXXVIII (5-6), 977-982.

Bellettati, D., Camassi, R. & Molin, D. 1993. Fake quakes in Italy through parametric catalogues and seismological complications: case histories and typologies. *Terra Nova* 5, 488-495.

Biers, W. R. 1992. *Art, Artefacts, and Chronology in Classical Archaeology*. Routledge, London.

Binford, L. 1982. Objectivity - Explanation - Archaeology 1981. In: *TAG 1980* (edited by Renfrew, C., Rowlands, M. & Abbott, B.). Academic Press, Southampton UK.

Blackman, D. J. 1973. Evidence of sea level changes in ancient harbours and coastal installations. , 115-137.

Blegen, C. W. 1926. The site of Opus. *American Journal of Archaeology (Second Series)* XXX, 401-404.

Bousquet, B. & Pechoux, P. Y. 1977. La seismicite du Basin egeen pendant l'antiquite. Methodologie et premiers resultats. *Bull. Soc. Geol. France* XIX(3), 679-684.

Bowen, A.,1998. The Place That Breached a Thousand ships. *The Classical Quarterly*, New Series. Vol. 48. No2. pp 345-364

Bowman, S. 1990. *Radiocarbon Dating*. British Museum, London.

Brumbaugh, D. S. 1999. *Earthquakes: Science and Society*. Prentice-Hall, New Jersey.

Buck, V., Stewart, I., 2000. A critical appraisal of the Classical texts and Archaeological evidence for classical earthquakes in the Atalanti Region, central mainland Greece. In: McGuire, W.J.,Griffths, D., Hancock, P.L., Stewart, I.S. (Eds.). *The Archaeology of Geological Catastrophes, Geological Society, London, Special Publications,* 171, 33-44.

Chrisman, N. R. 1997. *Exploring Geographical Information Systems*. John Wiley & Sons Inc, New York.

Christoskov, L., Gergova, D., Iliev, I. & Rizzo, V. 1995. Traces of Seismic effects on archaeological sites in Bulgaria. *Annali di Geofisica* XXXVIII(5-6), 907-918.

Clift, P. 1992. The collision tectonics of the southern Greek Neotethys. *Geologische Rundschau* 81 (3), 669-679.

Coleman, J. E. 1992a. Excavations at Alai in 1991. *American Journal of Archaeology* 96, 346.

Coleman, J. E. 1992b. Excavations at Halai, 1990 - 1991. *Hesperia* 61, 265-277.

Coleman, J. E. 1994. Halai, 1992-1993. *Archaeological Journal of America* 98, 287.

Cooper, F. A. 1978. *The Temple of Apollo at Bassai: A preliminary Study*. Garland Publishing, Inc, New York.

Cooper, F. A. 1996a. *The Temple of Apollo Bassitas*. The American School of Classical Studies at Athens, Princeton, New Jersey.

Cooper, F. A., Miller, S. G., Miller, S. G. & Smith, C. 1983. *The Temple of Zeus at Nemea: perspectives and prospects*. Benaki Museum, Athens.

Cooper, F. J. 1996b. *The Temple of Apollo Bassitas*. The American School of Classical Studies at Athens, Princeton, New Jersey.

Dakoronia, P. 1988. Αρχαικες Κεραμιδες Απο Την Ανατολικη Λοκριδα. *Hesperia* 59 (1), 175-180.

Dakoronia, P. 1993. Homeric towns in East Lokris: Problems with identification. *Hesperia* **62**(1), 115-127.

Dakoronia, P. 1996. Earthquakes of the Late Helladic III Period (12th century B.C) at Kynos (Livanates, central Greece). In: *Archaeoseismology* (edited by Stiros, S. & Jones, R.) Fitch Laboratory Occasional Paper No.7. British School at Athens, Athens, 41-44.

Dales, G. 1966. The decline of the Harappans. *Scientific American*, 92-100.

Davis, B. 1996. *GIS: A visual Approach (1ˢᵗ ed)*. Onword Press, Santa Fe

deMers, M.N. 1997. *Fundamentals of Geographic Information Systems* 1ˢᵗ Edition Wiley & Son

Demand, N. 1996. *A History of Ancient Greece*. McGraw-Hill, London.

Dermitazakis, M. D. 1984. *A guide to the Geology of Greece*. University of Athens, Athens.

Dickinson, O. 1994. *The Aegean Bronze Age*. Cambridge University Press, Cambridge.

DiVita, A. 1995. Archaeologists and Earthquakes:the case of 365 AD. *Annali Di Geofisica* **XXXVIII**(5-6), 971-976.

DiVita, A. 1996. Earthquakes and Civil life at Gortyn (Crete) in the period between Justinian and Constant II (6-7th century AD). In: *Archaeoseismology* (edited by Stiros, S. & Jones, R.). Fitch Laboratory Occasional Paper 7, Athens, 45-50.

Drews, R. 1993. *The End of the Bronze Age: Changes in Warfare and the catastrophe ca. 1200 BC*. Princetown University Press, New Jersey.

Dvorak, J. J. & Mastrolorenzo, G. 1991. The mechanisms of recent vertical crustal movements in Campi Flegrei caldera, southern Italy *Special Publication 263*. Geological Society of America, Colorado, 47.

Eagleton, T. 1996. *Literary Theory: An Introduction*. Blackwell, Oxford.

Evans, A. 1928. *The Palace of Minos at Knossos*. MacMillan & Co. Ltd, London.

Flemming, N. 1968. Holocene Earth Movements and Eustatic Sea Level Change in the Peloponnese. *Nature* **217**(5133), 1031-1032.

Flemming, N. 1969. Archaeological evidence for Eustatic Change in Sea Level and Earth Movements in the Western Mediterranean in the last 2,000 years. *Special Publication of the Geology Society of America* **109**, 125.

Flemming, N. C., Czartoryska, N. M. G. & Hunter, P. M. 1973. Archaeological evidence for eustatic and tectonic components of relative sea level change in the south Aegean. *Colston Papers* **23**, 1-136.

Flemming, N. C. & Webb, C. O. 1986. Tectonic and Eustatic Coastal Changes During the last 10,000 years Derived from Archaeological Data. *Z Geomorph. N. F* **Suppl. Bd 62**, 1-29.

Fossey, J. M. 1990. *The ancient topography of Opuntian Lokris*. Gieben, Amsterdm.

Frazer, J. G. 1913. *Pausanias's Description of Greece*. McMillan and Co. Ltd, London.

Gaffney, V. & Stancic, Z. 1991. *GIS Approach to Regional Analysis. A Case Study of the Island of Hvar*. Znanstveni Institut Filozofske Fakultete, Ljubljana.

Ganas, A. 1997. Fault Segmentation and Seismic Hazard Assessment in the Gulf of Evia Rift, Central Greece. Unpublished Doctoral Thesis, Reading, UK.

Ganas, A., Roberts, G. P. & Memou, T. in press. Segment boundaries, the 1894 ruptures and strain patterns along the Atalanti Fault, central Greece. *Journal of Geodynamics*.

Ganas, A., Wadge, G. & White, K. 1996. Fault segmentation and tectonic geomporphology in eastern central Greece from satallite data. In: *11th Thematic Conference and workshop on Applied Geologic Remote Sensing*, Las Vegas, Nevada.

Ganas, A. & White, K. 1996. Neotectonic fault segments and footwall geomorphology in eastern central Greece from Landsat TM data. *Special Publication of the Geological Society of Greece* **6**, 169-175.

Gergova, D., Iliev, I. & Rizzo, V. 1995. Evidence of a seismic event on Thracian tombs dated to the Hellanistic period (Sveshtari, Northern Bulgaria). *Annali di Geofisica* **XXXVIII**(5-6), 919-926.

Goldman, H. 1940. The Acropolis at Halae. *Hesperia* **IX**.

Goodchild, R. G. 1968. Earthquakes in Ancient Cyrenaica. In: *Geology and Archaeology of Northern Cyrenaica, Libya* (edited by Barr, F. T.). Holland-Breumelhof, Amsterdam, 215.

Guidoboni, E. 1989. I terremoti prima del Mille in Italia e nell'area Mediterranea (edited by SGA). SGA, Bologna, 765.

Guidoboni, E. 1996. Archaeology and Historical Seismology:the Need for Collaboration in the Mediterranean Area. In: *Archaeoseismology* (edited by Stiros, S. & Jones, R. E.) **Fitch Laboratory Occasional Paper 7**. British School at Athens, Athens, Greece, 7-13.

Guidoboni, E. & Binchi, S. S. 1995. Collapse and seismic collapse in archaeology: proposal for a thematic atlas. *Annali di Geofisica* **XXXVIII**(5-6), 101-1017.

Guidoboni, E., Comastri, A. & Traina, G. 1989. *Catalogue of ancient earthquakes in the Mediterranean area up to the 10th century*. Editrice Compositori, Rome.

Guidoboni, E. & Ferrri, G. 1989. The inexact catalogue: the study of more than 1700 earthquakes from the XI to the XX century in Italy. *Terra nova* **1**, 151-162.

Guidoboni, E., Riva, P., Petrini, V., Moretti, A. & Lombrdini, G. 1995. A structural analysis in seismic archaeoloy: the walls of Noto and the

1693 earthquake. *Annali di Geofisica* **XXXVIII**(5-6), 1001-1012.

Habicht, C. 1985. *Pausanias' guide to Ancient Greece*. University of California Press, London.

Hancock, P. & Altunel, E. 1997. Faulted Archaeological Relics at Hierapolis (Pamukkale), Turkey. *Journal of Geodynamics* 24(1-4), 21-36.

Hancock, P. L. 1986. Faulting. In: *Geology Today*, 15-151.

Hancock, P. L. 1987. Neotectonic fractures formed during extension at shallow crustal depths. *Momoir of the Geological Society of China* 9, 201-226.

Hancock, P. l. 1996. Ruptured Classical columns as recorders of past earthquakes. In: *Tectonics Study Group*, Cardiff.

Hancock, P. L. & Barka, A. A. 1987. Kinematic indicators on active normal faults in western Turkey. *Journal of Structural Geology* 9(5/6), 573-584.

Hope-Simpson, R. & Lazenby, J. F. 1970. *The catlogue of ships in Homer's Iliad*. Clarendon Press, Oxford.

Harris, T. & Lock, G. 1995 Towards and evaluation of GIS in archaeology: the past, present and future of theory and applications. In *Archaeology and Geographical Information Systems* (ed Lock. G & Stancic, Z. ) Taylor Francis. London

Hou, J.-J., Han, M.-K., Chai, B.-L. & Han, H.-Y. 1998. Geomorphological observations of active faults in the epicentral region of Huaxian large earthquake in 1556 in Shaanxi Province, China. *Journal of Structural Geology* 20 (5), 549-557.

Jackson, J. 1994. Active tectonics of the Aegean region. *Ann Rev Earth Planet Sci* 22, 239-271.

Jackson, J., Haines, J. & Holt, W. 1992. The horizontal velocity field in the deforming Aegean Sea region determined from the moment tensors of earthquakes. *Journal of Geophysicl Research* 97 (B12), 17,657-17,684.

Jackson, J., Haines, J. & Holt, W. 1994. A comparison of satellite laser ranging and seismicity data in the Aegean region. *Geophysical Research letters* 21(25), 2849-2852.

Kallner-Amiran, D. H. 1951. A Revised Earthquake-Catalogue of Palestine. *Israel Exploration Journal* 1, 223-246.

Kamberi, Z. 1993. Archaeological research and researchers in Albainia. *Institute of Archaeology Bulletin* 39, 1-27.

Karakhanian, A. S., Trifonov, V. G., Azizbekian, O. G. & Hondkarian, D. 1998. Relationship of late Quaternary tectonics and volcanism in the Khanarassar active fault zone, the Armenian Upland. *Terra Nova* 9(3), 131-134.

Karcz, I. & Kafri, U. 1975. Recent crustal movements along Mediterranean coastal plain of Israel. *Nature* 257, 297.

Karcz, I. & Kafri, U. 1978. Evaluation of supposed archaeoseismic damage in Israel. *Journal of Archaeological Science* 5, 237-253.

Karcz, I., Kafri, U. & Meshel, Z. 1977. Archaeological evidence for sub-recent seismic activity along the Dead Sea - Jordan Rift. *Nature* 269(15 September 1977), 234-235.

Kilian, K. 1996. Earthquakes and Archaeological Context at 13th C BC Tiryns. In: *Archaeoseismology* (edited by Stiros, S. & Jones, R.) Fitch Laboratory Occasional Paper 7. British School at Athens, Athens, Greece, 63 - 68.

Kissel, C. & Laj, C. 1988. The tertiary geodynamical evolution of the Aegean arc: A palaeomagnetic reconstruction. *Tectonophysics* **146**, 183-201.

Korjenkov, A. M. 1997. Study of the Nature and Mechanisms of Seismic For-runners of the Tien Shan Earthquakes. Laboratory of Seismological Methods of Earthquake Forecast, Kyrgyz Republic.

Lambeck, K. 1996. Sea-Level change and shore-line evolution in Aegean Greece since Upper Palaeolithic time. *Antiquity* **70**, 587-611.

LaRosa, V. 1995. A hypothesis on earthquakes and political power in Minoan Crete. *Annali di Geofisica* **XXXVIII** (5-6), 881891.

Lemeille, F. 1977. Etudes microtectoniques en Grece Centrale Nord-Orientale (Eubee Cenntrale, Attique, Beotie, Locride) et dans les Sporades du Nord (ile de Skiros). Unpublished These Docteur 3eme cycle thesis, Paris XI.

Levi, P. 1971. *Pausanias: Guide to Greece*. Penguin, London.

Lock. G & Stancic, Z. 1995. *Archaeology and Geographical Information Systems* Taylor Francis. London.

Lolling, H. G. 1876. Die Insel Atalande bei Opus. *Mittheilugen des Deutschen Archaeologischen Institutes in Athens*, 253-255.

Ma, X. Q. & Kusnir, N. J. 1995. Coseismic and postseismic subsurface displacement s and strains for a dip-slip normal fault in a three-layer elastic-gravitational medium. *Journal of Geophysical Research* **100**(B&), 12,813-12,828.

Ma, X. Q. & Kusznir, N. J. 1992. 3-d Subsurfacc displacement and strain fields for faults and fault arrays in a layered elastic half-space. *Gephysical Journal International* **111**, 542-558.

Manning, S. 1995. *Absolute Chronology of the Early Bronze Age*. Sheffield Academic Press, Sheffield.

Marco, S. 1997. 817 - year old walls offset sinistrally 2.1m by the Dead Sea transform, Israel. *Journal of Geodynamics*.

Marco, S. & Agnon, A. 1995. Prehistoric earthquake deformations near Masada, Dead Sea graben. *Geology* 23(8), 695-698.

Mariolakos, I. & Stiros, S. 1987. Quaternary deformation of the Isthmus and Gulf of Corinthos (Greece). *Geology* 15, 225-228.

Martin, T. 1996. *Ancient Greece: from prehistoric to Hellenistic times.* Yale University Press, London.

McCalpin, J. & Nelson, A. R. 1996. Introduction to Palaeoseismology. In: *Palaeoseismology* (edited by McCalpin) **62**. International Geophysics Series, London, 1-32.

McKenzie, D. 1972. Active tectonics of the Mediterranean Region. *Geophys. J. R Astr. Soc.* **30**, 109-185.

McKenzie, D. 1978. Active tectonics of the Alpine-Himalayan belt: the Aegean sea and surrounding regions. *Geophys. J. R. ast Soc.* **55**, 217-254.

McKenzie, D. & Jackson, J. 1986. A block model of distributed deformation by faulting. *Journal of the Geological Society* **143**, 349-353.

McKenzie, D. P. 1970. Plate tectonics of the Mediterranean region. *Nature* **226**(April 18), 239-243.

Miller, S. G. 1990. Nemea: A guide to the site and museum. University of California Press, Ltd, Oxford.

Miller, P. 1995. How to look good and influence people: thoughts on the design and interpretation of an archaeological GIS. In *Archaeology and Geographical Information Systems* (ed Lock. G & Stancic, Z. ) Taylor Francis. London

Mitsopoulos, K. 1895. *The Lokris Mega-earthquake,* Athens.

Morkot, R. 1996. *The Penguin Historical Atlas of Ancient Greece.* Penguin, London.

Mouyiaris, N. K. 1988. Destructive historical earthquakes in N.Euboicos and Maliacos Gulfs - Their significance to the evolution of the sea. In: *Proceedings of an international symposium organised by the Greek national group of IAEG* (edited by Marinos, P. G. & Koukis, G. C.). Balkema, Athens, 1249-1256.

Naumann, R. 1971. Wirkungen eines erdbebens an der antiken Bauten in Aezani. *Archäologisches Anzeiger* **86**, 214 - 221.

Negris, P. 1904. Vestiges antique submergees. *Athenischer Mitt.* **29**, 230-363.

Nikonov, A. A. 1988. On the methodology of archaeoseismic research into historical monuments. In: *Proceedings of an International Symposium organised by the Greek national group of IAEG* (edited by Marinos, P. G. & Koukis, G. C.). Balkema, Athens, 1315-132.

Nikonov, A. A. & Sildvee, H. 1991. Historical earthquakes in Eastonia and their seismotectonic. *Geophysica* **27**(1-2), 79-93.

Noller, J. S. & Lightfoot, K. G. 1997. An archaeoseismic approach and methodology for the study of active strike slip faults. *Geoarchaeology: An International Journal* **12**(2), 117 - 135.

Nur, A. & Ron, H. 1996. And the Walls Came Tumbling Down: Earthquake History in the Holy Land. In: *Archaeoseismology* (edited by Stiros, S. & Jones, R.) Fitch Laboratory Occasional Paper **7.**

British School at Athens, Athens, Greece, 75 - 85.

O'Hear, A. 1989. *An Introduction to the Philosophy of Science.* Oxford University Press, Oxford.

Oldfather, W. A. 1916a. History and topography of Lokris I. *American Journal of Archaeology* **XX**, 32-61.

Oldfather, W. A. 1916b. Studies in the history and topography of Lokris II. *American Journal of Archaeology* **XX**, 154-173.

Pantosti, D., de Martini, P. M.., Papanastassiou, D., Lemeille, F., Palyvos. N., & Stavrakakis. G.,2001. A reappraisal of the 1894 Atalanti Earthquake surface Ruptures, Central Greece. *Bulletin of the Seismological Society of America,* Vol 91, No. 4. pp 760-780.

Pantosti, D., de Martini, P. M.., Papanastassiou, D., Lemeille, F., Palyvos. N., & Stavrakakis. G., 2004. Palaeoseismological Trenching across the Atalanti Fault (Central Greece): Evidence for Ancestors of the 1894 Earthquake during the Middle Ages and Roman Times, *Bulletin of the Seismological Society of America,* Vol. 94, No. 2, pp531 – 549.

Papaconstantinou, M. P. 1996. A Seismic Destruction at Achinos (Phthiotis, Central Greece). In: *Archaeoseismology* (edited by Stiros, S. & Jones, R.) Fitch Laboratory Occasional Paper 7. British School at Athens, Athens, Greece, 87 - 91.

Papavsiliore, S. A. 1894. Sur le tremblement de terre de Locride (Grece) du mois d'avril. *Comptes Rendus* **119**(July-Dec 1894), 112-115.

Papazachos, B. & Papazachou, K. 1989. *The Earthquakes of Greece.* Ziti Editions, Thessaloniki.

Papazachos, B. & Papzachou, C. 1997. *The Earthquakes of Greece.* P Ziti & Co, Thessaloniki.

Philipson, A. 1951. *Der Nordosten Der Griechischen Halbinsel.* Vittorio Klostermann, Frankfurt.

Photinos, S. 1982. *Olympia Complete Guide.* Olympic Publications, Athens.

Pirazzoli, P. A. 1996. *Sea-Level Changes: The Last 20 000 Years.* John Wiley & Sons Ltd, Chichester, England.

Pirazzoli, P. A. 1998. A Comparison between Postglacial Isostatic Predictions and Late Holocene Sea-Level Field Data from Mediterranean and Iranian Coastal Areas. *GeoResearch Forum* **3-4**, 401-420.

Pirazzoli, P. A., Ausseil-Badie, J., Giresse, P., Hadjidaki, E. & Arnold, M. 1992. Historical environmental changes at Phalasarna harbor, west Crete. *Geoarchaeology: An International Journal* **7**(4), 371-392.

Pirazzoli, P. A., Laborel, J. & stiros, S. C. 1996. Earthquake clustering in the Eastern Mediterranean during historical times. *Journal of Geophysical Research* **101**(B3), 6083-6097.

Pirazzoli, P. A., Stiros, S. C., Arnold, M., Laborel, J., Laborel-Deguen, F. & Papageorgiou, S. 1994.

Episodic uplift deduced from Holocene shorelines in the Perachora Peninsular, Corinth area, Greece. *Tectonophysics* **229**, 201-209.

Rapp, G. 1986. Assessing archaeological evidence for seismic catastrophes. *Geoarchaeology: An International Journal* **1**(4), 365-379.

Rapp, G. & Gifford, J. 1982. *Troy: The archaeological Geology Supplementary Monograph* **4**. Princetown University Press, Cincinnati, 209.

Rapp, G. & Kraft, J. C. 1994. Holocene Coastal Change in Greece and Aegean Turkey. In: *Beyond the Site: Regional Studies in the Aegean Area* (edited by Kardulias, P. N.). University Press of America, Ltd., Maryland, USA, 69 - 89.

Reiter, L. 1990. *Earthquake Hazard Analysis: Issues and Insights*. Columbia University Press, Oxford.

Renfrew, C. & Bahn, P. 1991. *Archaeology :Theory, Methods and Practice*. Thames and Hudson Ltd., London.

Roberts, G. in press-a. Variation in fault-slip directions along active and segmented normal fault systems. *Journal of Structural Geology*.

Roberts, G. P. in press-b. Non-characteristic normal faulting surface ruptures from the Gulf of Corinth, Greece. .

Robertson, A. H. F. & Grasso, M. 1995. Overview of the late Tertiary-Recent tectonic and palaeo-environmental development of the Mediterranean region. *Terra Nova* **7**, 114-127.

Rondogianni, T. 1988. Recent tectonism and subsidence of historical sites at Lokris coasts, Greece. In: *Geologie de l'ingenieur appliquee aux travaux aciens, monuments et sites historiques* (edited by Marinos & Koukis). Balkema, Rotterdam, 1583-1589.

Rondogianni-Tsiambaou, T. 1984. Etude neotectoniques des rivages occidentaux du canal d'Atalanti (Greece centrale. Unpublished Doctoral thesis, Universite de Paris Sud.

Rotroff, S. I. & Oakley, J. H. 1992. Debris from a public dining place in the Athenian Agora. *Hesperia* **Supplement XXV**.

Schliemann, H. 1880. *Ilios: The City and Country of The Trojans*. John Murray, London.

Schliemann, H. 1884. *Troja*. John Murray, London.

Scholtz, C. H. 1990. *The Mechanics of Earthquake Faulting*. Cambridge University Press, Cambridge.

Schwartz, D. P. 1988. Geologic characterisation of seismic sources: moving into the 1990's. In: *Earthquake Engineering and Soil Dynamics II Proceedings*. US Geological Society, Park City, Utah, 1-42.

Schwartz, D. P. 1995. Earthquake repeatability: Magnitude, timing, segmentation. In: *ERICE*.

Shanks, M. & Tilley, C. 1987. *Social Theory and Archaeology*. Blackwell, Cambridge.

Skouphos, T. 1894. Die swei grossen Erdbeben in Lokris am 8/20 und 15/27 April 1894. *Zeitschrift Ges. Erdkunde zu Berlin* **24**, 409-474.

Smith, A. D., Andel, T. H. V. & Perissoratis, C. in prep. Late Quaternary sedimentation and sea-level/lake level changes in the North Evvioks Gulf, Greece. *Marine Geology*.

Sorren, D. 1981. Last Days of Kourion. In: *Studies in Cypriote Archaeology* (edited by Biers, J. C. & Soren, D.) Monograph XVIII. University of California, Los Angeles, 117 - 133.

Sorren, D. 1988. The day the world ended at Kourion: Recomstructing an Ancient earhquake. *National Geographic* **174**, 30 - 53.

Star, J & Estes, J.1990. *Geographic Information Systems: An Introduction*. Prentice Hall College Div

Stiros, S. 1986. Geodeticlly controlled taphrogenesis in back-arc environments: three examples from central and northern Greece. *Tectonophysics* **130**, 281-288.

Stiros, S. 1988a. Deformations of ancient constructions: Implications for the history of sites and seismotectonic research. In: *Engineering Geology of Ancient Works, Monuments and Historical Sites* (edited by Marinos & Koukis). Balkema, Rotterdam, 1591-1596.

Stiros, S. 1988b. Earthquake effects on ancient constructions. In: *New Aspects of Archaeological Science in Greece* (edited by Jones, R. E. & Catling, H. W.) Occasional Paper of the Fitch Laboratory Number 3. British School at Athens, Athens, 1-6.

Stiros, S. 1993a. Identification of historical and prehistorical earthquakes from their effects on ancient constructions. In: *Protection of Architectural Heritage against Earthquakes*, Istanbul, Ankara, 91-99.

Stiros, S. 1993b. Kinematics and deformation of central and south western Greece from historical triangulation data and implications for the active tectonics of the Aegean. *Tectonophysics* **22**, 283-30.

Stiros, S. 1995. Unexpected shock rocks an "aseismic" area. *EOS* **79** (50), 1513 & 519.

Stiros, S. 1996. Identification of Earthquakes form Archaeological data: Methodology, Criteria and Limitations. In: *Archaeoseismology* (edited by Stiros, S. & Jones, R. E.) Fitch Laboratory Occasional Paper 7. British School at Athens, Athens, 129-152.

Stiros, S. & Dakoronia, P. 1989. Rulo Storico e Identificazione di Antichi Terremoti nei siti della Grecia. In: *I Terremoti Prima del Mille in Italia e nell' area Mediterranea* (edited by Guidoboni, E.). SGA Storia-Geofisica-Ambiente, Bologna, 422-438.

Stiros, S. & Jones, R. 1996. Introduction/Editorial to Archaeoseismology. In: *Archaeoseismology* (edited by Stiros, S. & Jones, R.) Fitch Laboratory Occasional Paper 7. British School at Athens, Athens, 268.

Stiros, S. & Pirazzoli, P. A. 1995. Palaeoseismic studies in Greece: A review. *Quaternary International* **25**, 57-63.

Stiros, S., Pirazzoli, P. A., Laborel, J., Laborel, F., Arnold, M. & Papageorgiou, S. in press. Holocene sea-level changes in Euboea. *Bull Geol Soc Greece*.

Stiros, S. & Rondogianni, T. 1985. Recent vertical movements across the Atalandi fault zone (Central Greece). *Pageoph.* **123**, 832-848.

Stiros, S. C. 1985. Archaeological and geomorphic evidence of Late Holocene vertical motions in the N Euboean Gulf (Greece) and tectonic implications. Institute of Geology and Mineral Exploration (IGME), Athens.

Stiros, S. C. 1988c. Archaeology - A tool to study active tectonics. *EOS* **69**(50).

Stiros, S. C., Arnold, M., Pirazzoli, P. A., Laborel, J., Laborel, F. & Papgeorgiou, S. 1992. Historical coseismic uplift on Euboea Island, Greece. *Earth and Planetary Science Letters* **108**, 109-117.

Stiros, S. C., Pirazzoli, P., Rothaus, R., Papageorgiou, S., Laborel, J. & Arnold, M. 1996. On the date of construction of Lechaion, western harbour of Ancient Corinth, Greece. *Geoarchaeology: An International Journal* **11**(3), 251-263.

Taymaz, T., Jackson, J. & McKenzie, D. 1991. Active tectonics of the north and central Aegean Sea. *Geophysical Journal International* **106**, 433-490.

Thompson, T. F. 1970. Holocene Tectonic Activity in West Africa dated by archaeological methods. *Geological Society of America Bulletin* **81**, 3759-3764.

Trifonov, V. G. 1978. Late Quaternary tectonic movements of western and central Asia. *Geological Society of America Bulletin* **89**, 1059 - 1072.

Trigger, B. 1989. *A History of Archaeological Thought.* Cambridge University Press, Cambridge.

van Andel, T. H. 1989. Late quaternary sea-level changes and archaeology. *Antiquity* **63**, 733-745.

van Andel, T. H. 1994. Geo-archaeology and Archaeological Science. In: *Beyond the Site: Regional Studies in the Aegean Area* (edited by Nick-Kardulias, P.). University Press of America, Inc., Maryland, 25 - 44.

van Andel, T. H. & Runnels, C. R. 1988. An essay on the 'emergence of civilisation' in the Aegean world. *Antiquity* **62**, 234-47.

Vita-Finzi, C. 1978. *Archaeological Sites in their Setting.* Thames & Hudson, London.

Warren, P. & Hankey, V. 1989. *Aegean Bronze Age Chronology.* Bristol Classical Press, Bristol.

Wells, D. L. & Coppersmith, K. J. 1994. New empirical relationships among magnitude, rupture length, rupture width, rupture area, and surface displacement. *Bulletin of the Seismological Society of America* **84**(4), 974-1002.

Wren, P. 1996. Archaic Halai. Unpublished Masters of Arts Thesis, Cornell.

Yeats, R. S., Sieh, K. & Allen, C. R. 1997. *The Geology of Earthquakes.* Oxford University Press, Inc., Oxford.

Zangger, E. 1991. Prehistoric coastal environments in Greece: the vanished landscapes of Dimini Bay and Lake Lerna. *Journal of Field Archaeology* (18), 1-15.

www.ingramcontent.com/pod-product-compliance
Lightning Source LLC
Chambersburg PA
CBHW051302270326
41926CB00030B/4696